ASTRONOMY
and the imagination

The Author

Norman Davidson was born in Edinburgh in 1933. He was
a journalist for ten years, during which time he was a Fleet
Street correspondent for *The Scotsman* and also its drama
and film critic. For sixteen years following that he taught
astronomy, geometry, literature and history in Rudolf
Steiner schools and now lectures in general studies and
teacher training courses at the Waldorf Institute, Spring
Valley, New York. He is an amateur astronomer and a
member of the British Astronomical Association and the
Royal Astronomical Society of Canada.

J. Leslie White, the writer of the Foreword, is astronomy
correspondent of the *Daily Telegraph* and past President of
the British Astronomical Association.

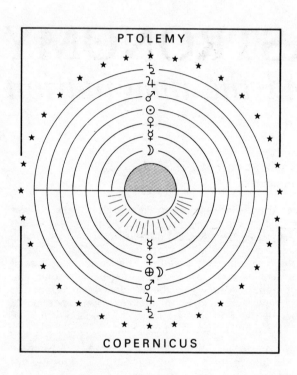

ASTRONOMY
and the imagination

A New Approach to Man's
Experience of the Stars

NORMAN DAVIDSON

Routledge & Kegan Paul
London and New York

First published in 1985
First published as a paperback in 1986
Reprinted 1987
by Routledge & Kegan Paul Ltd
11 New Fetter Lane, London EC4P 4EE

Published in the USA by
Routledge & Kegan Paul Inc.
in association with Methuen Inc.
29 West 35th St, New York NY 10001

Set in Linotron Palatino, 10/12pt
by Input Typesetting Ltd, London
and printed in Great Britain
by Billing & Sons Ltd
Worcester

Library of Congress Cataloging in Publication Data

Davidson, Norman

Astronomy and the imagination.
Bibliography: p.
Includes index.
1. Astronomy. I. Title.
OB43.2.D38 1985   520   84–15995

ISBN 0–7102–0371–3 (c)
      0–7102–1179–1 (p)

Contents

Foreword

At all times throughout history the human race has been interested in the skies of both day and night, and this interest has manifested itself in a variety of ways. For thousands of years before the invention of the telescope the changing aspects of the stars and the movements of the sun, moon and planets had been noted and recorded by learned men whom we rightly regard as astronomer-priests, for they believed that the destiny of the earth and mankind was intimately bound up with that of the celestial bodies. It is quite certain that astronomy would not have ceased to be a subject of interest at the end of the sixteenth century even if no new instruments were ever to be employed thereafter.

During the following three centuries there was an enormous increase in the accumulation of astronomical data and knowledge of the nature, motion and distribution of stars, all due to the ever-increasing means of investigation, commencing with the construction of bigger and bigger telescopes. The last four decades have seen a still greater expansion of astronomical research by ever more sophisticated technology, culminating in the truly staggering achievements of space probes to the planets involving the expenditure of unbelievably vast sums of money by several national governments.

Accompanying this more recent explosion in astronomical knowledge there has been a tremendous increase in the number of astronomy books published. A very large proportion of these are of a highly technical nature primarily intended for readers professionally engaged in astronomy and associated sciences. A great many others are popular accounts of what is being done by the new astro-technologists, presented in large, attractively produced volumes and lavishly illustrated with magnificent colour photographs of celestial objects and the amazing spacecraft and instrumentation with which they were obtained. What is often referred to as a great upsurge of public interest in astronomy is largely an interest in what is being done in these fields of highly specialised advanced technology, something very different from what was meant by astronomical interest up to the middle of this century.

There has also been, however, a very steady output of the more traditional kind of astronomical textbooks for students, from O level to degree courses, and of a wide range of popular introductory books, mainly descriptive of what can be seen with one's own eyes and simple telescopes and binoculars. It is these latter which are specifically directed to readers who have already actually looked at the night sky and have been impelled by a sense of wonder and curiosity to find out more about it. Many make astronomy their hobby, and some become serious amateur astronomers who engage in systematic observing programmes organised by local astronomical societies which have grown greatly in number during recent years.

The British Astronomical Association was founded in 1890 with the prime object of promoting amateur astronomy by the encouragement of its members with small telescopes to undertake systematic observational work under the guidance of the directors of the various sections devoted to the study of the sun, moon, planets and stars. For many decades it was the cherished belief of a large proportion of the membership that they were taking part in work of scientific value, a belief which is increasingly difficult to sustain in the context of present-day professional astronomy. It is certainly not held by hundreds of the current 3,000 members who derive great personal satisfaction from their own direct experience of the night skies in a wide diversity of astronomical occupations. They gaze at the celestial vista and *wonder*, as Man has done since time immemorial, seeking more than scientific explanation of phenomena solely in terms of scientific principles already educed in other fields.

It is for the wondering sky observer that this book has been written, a book quite different from any other among the present proliferation of astronomy books referred to earlier. It is fundamentally different in that, by its description and explanation of the constantly changing celestial scene, it emphasises at all stages that the watcher *experiences* a geocentric view of the heavens and that his knowledge of a heliocentric planetary system is acquired solely by inference. However assiduous his study of general and mathematical astronomy, he will never experience the sun-centred theory he has undoubtedly proved intellectually. His immediate knowledge of the objects beyond the earth is akin to his knowledge of what surrounds him on the earth, just as it has been for all people throughout the ages. In other words, heaven and earth are a totality of the experience of human life, something not realised by those who seldom look upwards.

This geocentric attitude to astronomy is admirably set forth at length in the introductory chapter to this book. The author considers in detail the whole question of how Man has interpreted what he has learned of his cosmic environment from ancient times right up into the space age. Many of those reading his comments will be delighted at finding this particular approach to astronomy which they will recognise as of special appeal to them, an approach which they will have found nowhere else. Others will be led to take a closer look at the sky than hitherto and be enthralled by their discovery of how much more there is to see and notice than they had previously realised.

Of still greater consequence, they will want to follow the progress of the various phenomena they have observed, not haphazardly but over weeks, months and years to witness from their own experience the completion of many of the cycles described in the book. They will also want to know the reasons for these systematic changes in the aspects of the stars and the cyclic returns to similar situations involving the planets. They will find simple explanations of the essential principles underlying the sometimes apparently complicated celestial happenings. Vital to these explanations are the excellent diagrams, many highly original, which form a large and indispensable part of the book.

Eclipses of the sun and moon are not common phenomena at any one place, and are often invisible because of clouds. Nevertheless, they are an intriguing subject about which people always want to know more than the circumstances of any particular one they may have seen. The reader will find far more information and explanation of these markers of history in Chapter 7 than he is likely to find in any other single book.

The concluding chapters on light in the sky and the telescope

are impressive examples of the whole intention of the book –
to urge people to *look* at the sky and to react to it with their
minds. My own experience over a great many years of teaching
astronomy in evening classes and of dealing with letters from
readers of the *Daily Telegraph* night-sky articles convinces me
that Norman Davidson's new approach to astronomy will
supply a very real need felt by many whose interest in the stars
prompts questions which require more than factual answers
from textbooks. The great, incomparable Camille Flammarion
still has no successor as a marvellous presenter of the wonder,
beauty and philosophy of astronomy along with its science, but
he has clearly been a source of inspiration to Mr Davidson, who
has written just the kind of book I have long wanted to write.

According to Bertrand Russell, Pythagoras was intellectually
one of the most important men that ever lived. Mr Davidson
has shown the way to the practical realisation of the truth of
the fundamental tenet of his philosophy: 'I am a child of Earth
and Starry Heaven.'

Leslie White, FRAS,
Astronomy correspondent of the *Daily Telegraph*

Acknowledgments

I wish to acknowledge the generous support of my colleagues at Michael Hall Rudolf Steiner School, Forest Row, Sussex, who gave the encouragement and practical opportunity for this book to be written.

I also wish to express my gratitude to Leslie White for offering many helpful suggestions at the draft manuscript stage and writing the Foreword following years of friendship and collaboration on astronomical matters of common interest; to John Alexander, Public Relations Officer at the Royal Greenwich Observatory, and other members of the staff, who showed so much patience and devotion to detail in answering my many questions; to Jean Meeus, the Belgian mathematician and writer on astronomy, who responded to my inquiries on celestial data so readily; to staff members of East Grinstead Library for their perseverance in searching out numerous unlikely titles, in a year of research, from shelves in various parts of the British Isles; and to all the friends and colleagues with whom I spent many days and nights debating which way the world goes round.

My thanks are due to the Husserl Archives in Louvain for permission to print a translation of extracts from Husserl's text referred to as 'Die Urarche Erde bewegt sich nicht' – the translation most kindly carried out by Dr Susan Arstall in collaboration with Professor Lothar Udert of Bochum University.

The frontispiece is adapted from a drawing by Giordano Bruno in his *La Cena de le Ceneri* (translated as *The Ash Wednesday Supper* by Stanley L. Jaki – Mouton, The Hague and Paris, 1975) which purports to reflect a discussion with Oxford scholars on the Copernican and Ptolemaic systems.

Gratitude for permission to use or adapt other illustrations goes to: the Lowell Observatory, Flagstaff, Arizona (Figure 12.5); Sidgwick & Jackson for material from Patrick Moore, *Year-*

book of Astronomy, 1984 (Figures 4.13 and 4.14); Cambridge University Press and Greenwood Publishers for material from Robert Greenler, *Rainbows, Halos, and Glories*, CUP, 1980 (Figures 11.5–11.9); Robert Powell for material from Robert Powell and Peter Treadgold, *The Sidereal Zodiac*, Anthroposophical Publications, 1979 (Figure 4.11); Oxford University Press and the British Academy for material from F. R. Hodson, ed., *The Place of Astronomy in the Ancient World*, 1974 (Figure 9.24a–d); John Mason for material from Patrick Moore and John Mason, *The Return of Halley's Comet*, Patrick Stephens, 1984 (figures 10.6, 10.7 and 10.8); The Huntington Library, San Marino, California, for material from S. K. Heninger, Jr, *The Cosmographical Glass*, The Huntington Library, 1977 (Endpiece: Eye, earth and cosmos in unity by Peter Apian); and acknowledgment also goes to Staatliche Museen zu Berlin for material from *Dürer – The Complete Engravings, Etchings and Woodcuts*, Thames & Hudson, 1965 (figure 2.8); Greenwood Publishers, Connecticut, for material from Robert Greenler, *Rainbows, Halos and Glories*, CUP, 1980 (figure 11.3); and the Smithsonian Institution, Washington D.C., for material from D. J. Warner, *The Sky Explored – Celestial Cartography 1500–1800*, Alan R. Liss and Theatrum Orbis Terrarum, 1979 (Appendix 4 figures A and B).

My sincere thanks also go to Ghislain Deridder, maths teacher at Michael Hall School, who assisted so readily in correcting the text for the second printing; and to Simon Wrigley of Luc Markies Associates (architects) of Forest Row, Sussex, who took such care in reworking some of the drawings for the second printing.

Introduction

Today there is a conspicuous lack of publications on astronomy dealing unashamedly with the sky as it is experienced by the ordinary, naked-eye observer standing under the stars. Science has become so sophisticated that a serious study of sky phenomena accessible to all is considered superfluous, outside textbooks and purely descriptive works on where to find the constellations, etc. Even in those, the earth-centred experience often quickly lifts off beyond the realm of the unaided eye into that of the large telescope, radio signals, and other up-to-date, and admirable, instrumentation.

Nowhere in print in English can I find a satisfactory publication on observational astronomy which treats it as a part of culture and as a creative science in itself. The present volume attempts to rectify this and is directed towards the ordinary member of the public who feels that the world of stars is a meaningful part of his greater environment and wishes to learn more about it within the bounds of his own experience; and it is also directed to the observational astronomer who requires a text which takes a fresh, intimate look at the movements and appearances revealed to his nightly vigil. In addition, the theme has a relevance to many aspects of culture including philosophy, psychology, history, literature and mythology. The aim has been to reintroduce the human cultural element into a basic science. Concordant with this is the inclusion, by modern authors, of the history of science and the evolution of consciousness into studies of, for example, history and literature. In this context the present book offers a contribution to an understanding of earlier cultures, permeated as they were with observational sky lore.

What it also offers, for a reader on any level, is one attempt at a fresh entry into astronomy arising out of Goethe's approach to science. A literary figure of considerable stature, Goethe (1749–1832) turned his comprehensive thinking and imagination to the sciences, including optics, which gave rise to his far-reaching and little-understood (if not misunderstood) Theory of Colours. His scientific writings were edited by Rudolf Steiner (1861–1925) whose further works and teachings resurrected Goethe's world view as the foundation for a renewal of culture. Steiner's *A Theory of Knowledge Implicit in Goethe's World Conception* and *Goethe the Scientist* constitute a deep scientific revolution which has not yet been fully realised and which has its roots in an inner need of the modern human being.

One requirement in astronomy is a study which is faithful to the observed phenomena, within a total human experience, allowing the phenomena to reveal themselves as a script. This is a creative process, the student being called on to make new beginnings, with everything relying on his own activity. The first chapter of the present text starts from an evaluation of the basic question of the geocentric and heliocentric systems. The sun-centred system and its derivatives hold the favoured position in science and the popular imagination. But the earth as centre is the archetypal system which works on in the unconscious despite Copernicus, because it is the cosmic ordering of 'direct impression'. In this sense it is the most 'effective' order. Dante's hierarchies and Jung's synchronicities have their dwelling there.

The geocentric approach seems much misunderstood and misrepresented in literature today. It is not realised how much it plays into our earthly existence. (Husserl's text, quoted in Appendix 1, investigates the experience and concept of movement from a geocentric position.) For the earth inhabitant, the earth is the centre and is not experienced as moving, but this does not preclude a moving earth as understood from some other position or viewpoint. For someone living on Mars, the universe is Mars-centric. While citizens of earth, we cannot ignore the geocentric view – or else cannot escape it even if we ignore it. I have sat in a planetarium and been lectured on the obsolete nature of the illusory geocentric view and how modern man has outgrown it, but the hapless lecturer could not avoid pressing a perfectly modern button to darken the auditorium and turn the night sky round the central audience. Phenomenologically his action was entirely correct.

One question presenting itself to the student of astronomy with an eye for qualitative appearance, is what department of science astronomy stands within. In one sense it is inorganic and mathematical and in another it is, in its wholeness, organic and living. Thomas Aquinas in *The Division and Methods of the Sciences* (a commentary on the *De Trinitate* of Boethius) placed it among the 'intermediate' studies, along with music and optics, between mathematics and natural science. Certain sciences he called 'arts because they involve not only knowledge but also a work that is directly a product of reason itself' – such as discourse, composing melodies or 'reckoning the course of the stars'.

It seems to me that astronomy can relate like a mother-study to all the sciences and to many other aspects of human life. In the form presented here, it has been sadly neglected in recent times and needs to be reinstated into education and general culture. Much is said and written today on cosmologies and star lore past and present. For this to be fully effective it requires a firm foundation in the experience of the circling stars.

The 'imagination' of the title of this book must not be confused with any fanciful, other-worldly treatment of the subject. Another author may use another word, but here it points to a vital and integral part of human consciousness which is today too often ignored and, when it is, inevitably reasserts itself – but then emerges as a distortion leading into the realm of fancy, loose association and mere invention. Thus, nature is not to be denied and takes its revenge, for we end up with mental creations which have no direct connection with the actual working of things.

If a Chinaman who always uses the floor to sit on looks at a chair for the first time without imagination, then it is to him just an incongruous fabrication of wood and glue. He may use it as a table, but he will not penetrate to its origin and meaning – to the thought behind it and the effect it has for human life.

In his *Biographia Literaria* Coleridge makes a clear distinction between imagination and fancy. His idea of a primary form of imagination is to see it as 'the living power and prime agent of all human perception'. A secondary form of imagination 'struggles to idealise and to unify. It is essentially *vital*, even as all objects (as objects) are essentially fixed and dead.' He adds,

'Fancy, on the contrary, has no other counters to play with but fixities and definites . . . equally with the ordinary memory it must receive all its materials ready made from the law of association.'

Fancy works upon, and is led by, the exteriors of things but imagination penetrates phenomena, bringing them alive. Fancy brings something to objects from outside, whereas imagination, at root, is the 'prime agent of all human perception' and unites mind and object in the act of understanding, and of discovering meaning.

Goethe once wrote to the philosopher Friedrich Jacobi: 'God has punished you with metaphysics and put an arrow in your flesh; me He has blessed with physics . . . You hold fast to faith in God, I to beholding.' Goethe was a poet and what he meant by physics and beholding was no dry rendering of these words. His beholding was active and imaginative and was wedded faithfully to the world of phenomena. In this way the power of imagination finds its home in the realm of direct experience. Such is the concept and role of imagination adhered to in this book, taking its initial cue from Coleridge and Goethe.

Norman Davidson, Forest Row, 1984

[T]he sight of day and night, the months and returning years, the equinoxes and solstices, has caused the invention of number, given us the notion of time, and made us inquire into the nature of the universe; thence we have derived philosophy, the greatest gift the gods have ever given or will give to mortals. This is what I call the greatest good our eyes give us.

PLATO

Who could deny the sacrilege of grasping an unwilling heaven, enslaving it, as it were, in its own domain, and fetching it to earth?

MANILIUS

I know that I am mortal and ephemeral, but when I scan the crowded circling spirals of the stars I do no longer touch the earth with my feet, but side by side with Zeus I take my fill of ambrosia, the food of the gods.

PTOLEMY

I know not what the world will think of my labours, but to myself it seems that I have been but as a child playing on the sea-shore; now finding some pebble rather more polished, and now some shell rather more agreeably variegated than another, while the immense ocean of truth extended itself unexplored before me.

NEWTON (in old age)

The paramount thing would be to comprehend that everything factual is already theory. The blue of the heavens reveals to us the basic law of chromatics. Only let one not seek for something behind the phenomena; they themselves are the teaching.

GOETHE

The best thing in the sciences is their philosophical
ingredient – like life in an organic body.
Dephilosophise the sciences and what is left?
Earth, air, and water.

NOVALIS

The primary imagination I hold to be the living
power and prime agent of all human perception,
and as a repetition in the finite mind of the eternal
act of creation in the infinite I AM.

COLERIDGE

Poetry is indeed something divine. It is at once the
centre and circumference of knowledge; it is that
which comprehends all science, and that to which
all science must be referred.

SHELLEY

If we learn the writing spread forth in the cosmos,
in the stars, in their ordering and motions, we shall
find that out from the cosmos everywhere speaks
that which permeates our hearts with truth, love,
and that piety which carries forward the evolution
of humanity from epoch to epoch.

RUDOLF STEINER

Every phenomenon can be experienced in two ways.
These two ways are not arbitrary, but are bound up
with the pheneomenon – developing out of its
nature and characteristics: Externally – or –
inwardly.

KANDINSKY

[S]cience must rely upon ordinary language as well,
because this is the only language in which we can
be certain to get hold of the phenomena . . . If
harmony in a society depends on the common
interpretation of the 'one', of the unity behind the
multitude of phenomena, the language of the
poets may be more important than that of the
scientists.

WERNER HEISENBERG

Chapter 1
The Earth as a Centre

This book attempts to make a new beginning in astronomy. It supports the case for an earth-centred view but this neither means a mere reversion to an old concept nor a denial of modern research into a sun-centred or galaxy-centred or non-centred universe. These recent advances are realities, but it must be properly assessed what kinds of realities they are. They are not founded on the direct experience of the ordinary human faculties of sight and the senses of space and movement natural to the human being in his normal, earth-based environment. They are removed from this with the help of technology and have a different validity. A study of these modern developments and their significance would fill another volume.

The attempt here is therefore not to deny modern research but to encourage a process of placing matters in perspective, which may eventually lead to a reassessment of this research in terms of human experience.

Such reassessment is largely beyond the scope of this present text which simply explores a new beginning from first principles. In histories of astronomy or in basic books describing the subject, the first chapter or so is often given over to 'ancient' earth-centred theories in a condescending way, seeing such theories as the result of ignorance or the child-like concepts of a people unenlightened by the discoveries of modern science. It is considered naive today to devote a whole book to a study of celestial phenomena rooted in the direct experience of the ordinary observer. But the neglect of this area has resulted in a serious loss of the human element in this earliest and most fundamental of sciences. The real motives of earlier cultures in adhering to an earth-centred view are not fully appreciated. The earth-centred view engendered a connection in feeling between the individual and the universe; Man was part of the whole, an idea which, since Copernicus, is considered out-moded. But there is research to be done in continuing a study of modern man's relationship to the universe from the aspect of direct, personal experience.

It is even said that it is baffling why the Greeks persisted in an earth-centred view even after Aristarchus of Samos (c. 310–264 BC) declared that the earth moved round the sun with the other planets. But the Greeks had an innate feeling that Man was connected directly with the whole universe, and a sun-centred concept would have been pointless for the majority of them. Traces of this feeling remained around the time of Copernicus in the Church's resistance to the sun standing still. It was not only a case of the Church authorities resting on tradition – although that was also part of the picture.

The Copernican Revolution had to come. It is one of the greatest achievements of the human intellect. But that does not mean it is the only way to see reality. Since Copernicus there have been immense strides in physical science. But physical science and the intellect do not make up the sum total of human nature, even though they are an important part of it. They do not necessarily arrive at the meaning of phenomena for the individual. As Goethe once said, nature is like a woman – 'She delights in illusion. Who kills illusion in himself or others, him she will punish as the sternest tyrant.' Put in another way, we can intellectually and physically analyse what is female, but this, although one form of reality, will never explain a man's love and why 'she' means everything to him, inspiring wider horizons in human nature. Yet this is a reality of everyday existence. If it is not included in the picture then the picture must be one-sided and our connection with life diminished.

The mistake is in imagining that two fundamental sides of reality contradict each other; the sun-centred view and the earth-centred originate from different aspects of the one situation, but they do not contradict. One supplies the needs of abstract, material thinking, the other of direct human experience. There is no need to choose between the two, though one or the other will be uppermost depending on what we are looking for.

The Copernican Revolution, resulting from the idea that the earth moved, brought about the greatest psychological change in human consciousness for many centuries. It had its effects in all areas of life and changed society. It makes a huge psychological difference whether one believes one is living on an earth which is moving in space round a centre outside it, or whether one believes one is living on an earth at rest with the universe moving round it. Copernicus's idea, launched in the first half of the sixteenth century, could not be established satisfactorily until the first part of the eighteenth century.* Between these times the movement of the earth was conjectured and calculated by many astronomers who were already convinced of the fact, and the general public followed in their wake. The time was ripe for the change.

Man's view of the universe he lived in was now based on calculation and subtle scientific instruments, both of which were beyond the experience of the man in the street. Even stronger, therefore, could the new world view be taken up by the general public in rational thinking which could thrive on just such a sense-free notion and lead at the same time to a material concept of the universe. This material concept was expanded by the development of the telescope and perceiving planets as physical bodies with shadows. This in turn led to the idea that human life was microscopic in relation to the stars and almost incidental.

With Copernicus's great innovation and the discovery of the telescope, the lay person was, in fact, less likely to look up at the stars and follow their movements. This latter activity was unnecessary as he was looking at an 'illusion'. He 'knew' that the sun, planets and stars did not rise and set of themselves, but that the earth turned. But he only knew it and did not experience it. Not even the scientist, in fact, experiences it in a direct way. For the ordinary observer on earth the sun is seen to rise. He does not see or experience the earth turning. But he can think the earth to be turning. If he takes the modern planetary system entirely seriously he must think a lot more – for instance, in which direction in space the earth is moving, from where he stands, at any particular time in its passage round the sun, something which few people would consider when looking at a sunrise. The modern system of planetary movement is, in fact, based on a hypothetical observer sitting in space far enough away from the planets to enable him to view them, within his immediate field of vision, going round the sun. The fact is that if he were placed thus far enough away he would not see the planets at all. No one has ever seen or ever will see the sun-centred system as drawn in books.

This typifies a fundamental division between sense experience and thought which lies unnoticed at the root of our modern condition. It is the divorce of thinking from living experience, which results in the domination of an abstract thinking. Almost without realising it we are in the middle of a historical phase in which the human being has a split between his intellect and his feeling for the world around him. The intellect, which is an important faculty, has nevertheless grown to tyrannise the life of feeling and imagination in their relation to phenomena. The result is that imagination either withers or takes on fantastic forms of its own (there is plenty of science fiction in science today) divorced from experienced reality. We must use all human faculties to achieve a balanced understanding of the world, and intellect and imagination should work together. Otherwise the intellect by itself will create a cold, mechanical world for us, as equally divorced from reality as fantasies.

The science historian Giorgio Santillana referred to the psychological split caused by modern astronomy in *Hamlet's Mill* when he said:

When [Man] discovers remote galaxies by the million, and then those quasi-stellar radio sources billions of light-years away which confound his speculation, he is happy that he can reach out to those depths. But he pays a terrible price for his achievement. The science of astrophysics reaches out on a grander and grander scale without losing its footing. Man as man cannot do this. In the depths of space he loses himself and all notion of his significance. He is unable to fit himself into the concepts of today's astrophysics short of schizophrenia.

We are often obsessed with 'explaining' phenomena rather than experiencing them humanly also. For the universe is meaningless if not related to the human being; space becomes a void, dispossessed of a centre and is therefore a concept wherein Man cannot find a 'place'. He is nowhere, and without, for example, his dynamic sense of direction in 'up' and 'down', 'left' and 'right', etc.

The contention of this book is that direct human experience should not be neglected or passed over quickly and that, in fact, it must come first. The public is showing renewed interest in this direction and the balance may already be swinging back – though the spread of pseudo-occult scientific literature which simply mystifies on the one hand and makes artificial connections on the other only prevents any healthy change.

There is a common conception that the earth-centred view of earlier cultures was not only simple-minded but that it was egotistic as well. Putting Man at the centre of the universe is considered to be too special a position for him, inflating his sense of importance. Yet taking him away from that centre and denying its part in his experience has led to a disconnection between the individual and the universe and the loss of a sense of purpose in relation to it. In *A Sense of the Cosmos* Jacob Needleman points out that

> [I]n ancient geocentricism the spheres or forces that
> surround the earth are both more powerful and more
> subtle than anything originating on the earth itself.
> Understood in this way, geocentricism humbles man
> and calls him to search for a finer understanding of the
> influences that shape his life and the life of the world.
> It is therefore a great mistake to assume, as all modern
> writers have, that ancient geocentricism exaggerated
> man's importance in the scheme of things. For to be at
> the center meant in effect to be at the lowest rung in
> the ladder of influences.

Then later:

> But taken with the idea of the microcosm, geocentricism
> reminds man that objective reality contains many kinds
> of influences that can act upon us, that there is a scale
> of being to which man is born would he but search for
> it as diligently as he pursues the satisfactions of external
> life.

Having a centre where oneself as a human being is situated, is a normal experience – in fact one is abnormal if the experience is diminished or absent. Sensing a periphery of influence in the shape of one's environment is also a part of normal experience. Absence of a sense of this polarity between centre and periphery and their interaction leads to imbalance and disorientation. The healthy life of the individual lies between the two.

Though the movement of the earth may be calculable it is not, as put before, a direct sense experience. A preliminary development of this fact from the philosophical standpoint was made by Edmund Husserl in 'The Earth Does Not Move', a short text written in 1934, extracts from which are included here in Appendix 1. Pure mathematics and calculation are indifferent to the quality or import of phenomena. Sun-centred astronomy is a mathematician's astronomy. But the geocentric experience unites earth, man and sky into a whole. This experience should not be denied, because it is a direct connection with the environment as it affects us and this is a primary not a secondary reality. Obvious effects are day and night, the seasons, the moon – the various recognisable rhythms of life. Others may pass unnoticed or unrecognised or be explained into silence. For example, merely to explain that Jupiter makes no loop against the stars but instead the earth overtakes it once a year, and to leave the matter at that, gives rise to the thought that the loop is only an illusion, not a phenomenon of nature, and therefore not to be taken seriously. But if a professor were to give his lecture while apparently striding around in loops all the while, this would have a direct effect on the students' nerves despite any explanation that the students were in orbit or that it was done with mirrors. The reality for the students is that the professor is walking in loops and that they feel ill watching him.

To express it in another way, the night sky is a theatre within which the dramas and events of the universal environment take place. To explain that, when not seen on stage, Romeo and Juliet are not interested in each other at all and are millions of light-years apart emotionally; or that the pomegranate tree in Capulet's orchard is oil paint and plaster, is secondary to the effect of the play on the audience.

Not that inquiry into what the telescope and microscope show is unimportant. It is part of our modern experience in the science of matter. Indeed, the development from ancient to modern cosmology reveals the evolution of human consciousness. But modern research must not separate itself off and stand apart from a contemplation of the whole which starts with the observed phenomena in their simplicity and retains a connection with the aesthetic, qualitative appreciation of life. For this latter purpose a faithful study accessible to ordinary, unsophisticated observation is required. Then a living connection with nature is felt which employs the complete human being.

This question of uniting the human being with the phenomena can be taken a stage further in observational astronomy. This involves the reintroduction of mythology into the experience of the turning sky. If one is honest, it is very difficult to keep mythology out. The subject is dismissed as superstition today, yet we replace it happily with mythologies about curved space, civilisations on other planets, black holes, etc., none of which have been experienced. The human being must provide a meaning or content to the universe, otherwise there is a gap left in his thinking. In earlier times the stars were looked at personally and were felt to be active participants in the drama of life. Mythology readily springs out of geocentric astronomy and strengthens the connection with the phenomena. The emergence of mythology cannot be an arbitrary process, but has its roots in the quality of what is experienced. For example, the ancient designation of metals to the natures of the planets can be understood by a study of the planets' characteristic movements, e.g. the slow heaviness of Saturn and lead; the speed and liveliness of Mercury and quicksilver. Likewise is the connection between the unmoving, central position of the Pole Star and heavenly authority, etc. It is well nigh impossible to shake free of unifying concepts of one sort or another when contemplating the sky as it appears – such concepts are naturally called forth in human experience, even if the concepts are modern ones. So mythology emerges as part of this present book of observed phenomena, not because the mythologies add some irrelevant interest or should be unthinkingly believed, but because they form a natural part of the human activity of watching the stars shine. Many mythologies are possible and only a few have been selected. But the important thing is that the activity of myth-making is recognised honestly and, for the future, that new mythologies arise as human evolution develops – mythologies or metaphors or star pictures transcending the current scales of distances in light-years.

Two ways of observing phenomena emerge from this discussion. There is the dualistic approach which observes the phenomena then takes a separated, removed position and imposes an explanation by means of other, unobserved factors (mathematics for example) or indirectly observed factors (atomic physics, etc.). Then there is the integrated approach which observes the phenomena and enters them with the whole human being involved, identifying with them and waking up within them, so to speak. The total environment is not lost and the signature of the phenomena and their relation to life can be read.

Arising from this latter approach in astronomy, which starts from the geocentric view, other astronomies can develop, be they heliocentric or whatever. But they must be experienced and not just abstractly thought. For instance, if the movement of the earth became an experience, not just an abstract thought, this could generate an entirely new astronomy, not necessarily heliocentric; and it would remain part of the human being.

As for the geocentric starting point, nature responds to eclipses, the apparent movements of the sun and moon, etc., and the human being can also. It is in his essential nature to do so. If true to himself he will, out of his freedom, befriend the appearance of the stars and let the phenomena, which make up his greater environment, declare themselves.

* In 1729 James Bradley announced his telescopic discovery of the aberration of light which accounted for an apparent displacement of star positions close to overhead as being due to the orbital movement of the earth. In 1838 Friedrich Bessel correctly established the annual parallax of a star (a faint nearby one in the constellation of the Swan) which directly showed, by minute telescopic observations, the earth's movement reflected in the star's apparent movement.

Chapter 2
Seeing the Phenomena

The essential thing which astronomy requires of the human being is for him to step outside, look upwards and consider his widest environment. This is no trivial act. It achieves, principally, two things.

Firstly, ordinary thinking is given over to the contemplation of something patently beyond, and greater than, itself. There is a refreshing release and expansion into a realm at once objective and awesome in its beauty and immensity. What professional astronomer, with his instruments and books of calculations, or lay citizen has not looked on the face of the starry sky and been held by a feeling of natural wonder which arouses questions of life's very mystery?

Secondly, after study of the movements and appearances of what confronts the observer, it is realised that the human being is, consciously or unconsciously, deeply bound up with all the celestial phenomena of time and rhythm. More than 2,000 years ago Plato, in his dialogue *Timaeus*, went so far as to say that the purpose of the gift of sight was that we should see the movements and rhythms of the heavens and thus learn how to put right those within ourselves. It is noteworthy that astronomy is differentiated from the other sciences by relying essentially on the faculty of sight for its investigations. At root, astronomy is the perception of lights in motion around us, and even the most physical inquiries into its secrets start from this.

For the earth-centred observer three things are borne in on him which form the basis of all that he then develops. Firstly, that the sky above him is experienced and directly thought of as a dome or hemisphere traced out by the stars; secondly, that this dome meets the plane of the earth's surface in a line or horizon; thirdly, that the observer is a centre point to this

9

picture. Hemisphere, plane, line and point comprise the theatre for celestial events to happen (Figure 2.1). This geometrical model is a concept living in the imagination, but it still remains a part of the immediate experience.

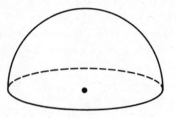

Figure 2.1

Seen from the observation point in any one direction, the line of the horizon, say at sea, is level, so it can be represented by a straight line – though it gives a slightly curved appearance due to its being concave to us. Above it an approximate arc of vision passes overhead from side to side (Figure 2.2). This arc is not a semi-circle, but is flattened at the top, as one sees further in any one moment from side to side than upwards.

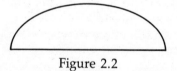

Figure 2.2

If one then sweeps one's gaze round the sky, an impression is of standing at the centre of a sphere, half of which is visible.

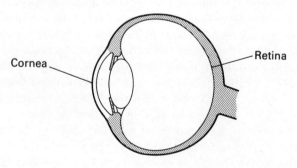

Cornea   Retina

Figure 2.3

This arises from the direct, phenomenological experience of the eye and of our mental imaging. The shape of the eye itself is modelled on a sphere and the technical name given to the eye socket is the orbit. Both the outer front of the eye (cornea) and the larger part of the inside which receives the image (retina) are related to the sphere (Figure 2.3).

How the eye is not merely a mechanical camera is evidenced by the fact that the image received by the retina is inverted, yet we see the 'right' way up. As M. H. Pirenne points out in *Optics, Painting and Photography* –

> Vision is an active, not a passive process. In order to see it is necessary to look . . . The old theory of emission of visual rays gave a direct suggestion of the active part played by the observer in visual perception. In the modern theory this might appear to be left out of account.

He quotes the writer on optics Le Grand as saying that the eye is the only optical instrument which forms an image which has never been intended to be seen.

It can be made to appear that the eye, being separated from the world around, stands apart as spectator. But it is not the eye itself which sees, it is only the vehicle to seeing. Therefore it is moulded on the archetype of the greater visual environment – the sphere, plus light and darkness. It is an interesting fact that the reason for the blue appearance of the day-time sky is the same as that for the blue in the human eye. Both are not physical colours or pigments, but are colour effects arising out of light and darkness. The blue sky is caused by the darkness of outer space being overlaid by the lit, semi-transparent substance of the atmosphere. The black background is modified into blue by the lit substance standing between the observer and the background. The same principle holds for the iris of the eye which, in blue-eyed people, has a dark backing overlaid with a layer of thin, white substance, producing colour. The pupil of the eye, having no such white surface substance, looks as dark as night.

The principle of the eye and of the celestial vault cannot be separated in the act of seeing. It seems no accident that Johannes Kepler, one of the greatest of astronomers, researched the eye and did for optics what he did for astronomy by laying the foundations for modern science in both these fields. Kepler was the first to understand clearly that an inverted image is formed on the human retina, as in a pinhole camera or camera obscura.

We can now take the next step into seeing what appears to lie upon the celestial sphere. The observer stands at the centre on the plane of his horizon (Figure 2.4). The celestial sphere of

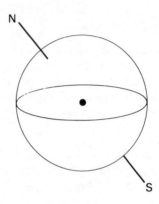

Figure 2.4

stars then appears to rotate round an axis, directed due north and south. The turning of the sphere is from east to west, explained as reflecting the rotation of the earth from west to east. To help in determining the positions of stars on the sphere we can again look at it from the outside and extend the plane of the earth's equator out to meet it in a circle called the celestial equator. It can then be conceived that any star on it moves in a day on a circle whose plane is parallel to the celestial equator. Let a broken line represent the celestial equator, which is an extension of the earth's equator, and it will pass from a point directly east on the horizon, curve upwards to the south and meet the horizon again due west. The plane of the celestial equator will, of course, be at right angles to the line of axis through the poles (Figure 2.5). If we add the daily paths of stars, then they will appear to move in circles whose planes are parallel to that of the celestial equator (Figure 2.6). This is how we draw the situation from outside and when we stand on the earth looking around and upwards at the sky we have the experience of the same thing from within the sphere.

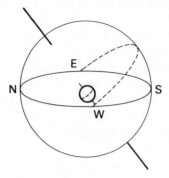

Figure 2.5

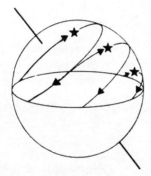

Figure 2.6

Yet it is important to realise that geometrically there is no way of reproducing on a plane a picture of star positions exactly as they appear on a sphere. Every star map must be based on one of many methods of projection onto a flat plane, which plane necessarily distorts the positions which are being plotted from a sphere. A photograph of the night sky is a good example of this, for the camera can never tell the truth about the world we see. An ordinary photograph of stars projects a fixed image through one point of perspective onto the flat plane of the film and, eventually, onto the page. But the eyes are neither fixed nor have only one point of perspective nor project an image onto a flat plane. We have two eyes which move when we look at a scene, which focus from different points of perspective with every movement, and which project the image onto their spherical shape.

Despite this, one is familiar with diagrams and photographs in books on astronomy which do not explain this. Instead, for instance, photographs are reproduced after pointing the camera, say, in an easterly direction, on long exposure and showing the resultant star 'trails' as curves which lie on either side of a straight line representing the path of a star rising due east (Figure 2.7). Some photographs or diagrams showing this

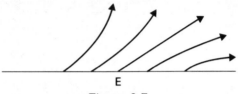

E

Figure 2.7

also state that these curves are the movements seen by the observer from his perspective position at the centre of the celestial sphere. But this is only so for the camera, or for a

Figure 2.8 Drawing a lute in perspective with pointer and string (woodcut by Albrecht Dürer)

projection onto a flat plane. This photographic type of representation was pioneered at the time of the Renaissance by artists
who developed linear perspective (Figure 2.8), but it was
realised that this geometrical method was far from representing
correctly what is seen. A struggle then began, which still
continues, to break up this mathematical, photographic-type
perspective into a subtler method which came closer to what
the eye sees (the subtler method has been called synthetic
perspective). Leonardo da Vinci experimented with both types
of perspective. In the mathematical type he stated that 'Perspective is nothing else than seeing a place or objects behind a pane
of glass, quite transparent, on the surface of which the objects
which lie behind the glass are to be drawn.' Applied to moving
stars seen through a window, this would produce the type of
curves shown in Figure 2.7. But he also tried to develop a
synthetic perspective in the plane which was more faithful to
the eye's 'spherical' experience and based it on the rays of light
between object and eye being intersected geometrically by a
sphere.

The problem is summed up by Pirenne, quoted earlier, who
says:

> Retinal images occur as one link in the chain of events
> which constitutes the process of seeing. It is not this
> link that pictures, photographic or otherwise, are
> intended to duplicate: it is the external visible world
> itself. The purpose of photographs is to be seen. It is not
> the purpose of the photographic camera to see.
> Consequently there is no reason why photographs
> should mimic the peculiarities of the retinal image . . .
> While photographs are intended to be looked at, retinal
> images are not. The photographic camera is not an eye.

A study of astronomy should not only be linked with external
optics but with the inner activity of the observer. What we
'passively' see and what we actively think about what we see
are different things. For instance, we can never actually see two
parallel lines as measurably equidistant along their length from
where we stand, but because of our position in relation to them
we always see them as ultimately converging to a point – as
when we look up a railway line. The rails we see as converging,
but in doing so we also 'think' of them as being parallel, other-

wise they would have no meaning for us. When we pick up a cup we don't imagine that the rim in itself begins as elliptical when the cup is on the table and changes its shape to become more circular as we tip it up to drink. The percept and the concept are united in one single experience. Other situations are more abstract and lead to more separated concepts, as in observing sunbeams below a cloud and 'seeing' them radiate from a point (the sun) behind, or in observing a radiant of meteors issuing from a point in the sky. Both sunbeams and meteors, in the light of further reasoning, are virtually parallel.

Even in the case of the united concept and percept, seeing is active, requiring the thinking and understanding of the observer, as the psychologist will admit. Yet, apart from writers like Pirenne, science today rejects the descriptions of Plato, Euclid and other early thinkers of 'seeing' involving light, of a kind, streaming from the eye of the beholder and blending with the light emanating from the object. This light in the eye could be understood as the light of thinking, of active concept-forming.

To return to star 'trails', one appropriate way of representing the movements of stars on the east and west sides of the celestial sphere is to draw their paths in parallel straight lines. When standing under the sky, all star paths, in fact, appear to move 'parallel' to each other on curves which are experienced as circles as a star always appears the same distance away from the observer. The picture seen in eastward and westward directions above the horizon in mid-northern latitudes could be shown as in Figures 2.9 and 2.10.

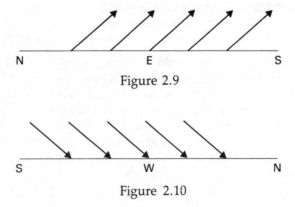

Figure 2.9

Figure 2.10

If, however, the north and south points of the horizon are included in the drawings, as shown, then this represents something which the observer cannot perceive clearly within his field of vision while looking in one direction. Therefore such a drawing assumes that the observer turns left and right to take in further points on the level horizon. Consequently, the south and north sides of the horizon appear as in Figures 2.11 and 2.12. On the plane of the page, these are oval-type curves, here related partially to height above the horizon and distance along it. We stand, of course, within the plane of one of them, the celestial equator, which passes from due east to due west (Figure 2.11) but the impression to the picturing eye under the stars is not of a straight line but of a line curved concavely to us, and to which other movements in the south are 'parallel'.

The degree of curve should alter with the latitude on earth of the observer, and is represented here as circular when

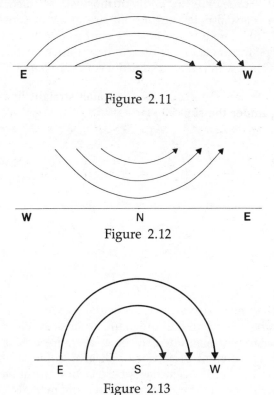

Figure 2.11

Figure 2.12

Figure 2.13

looking south from the equator (Figure 2.13) and flattening to straight lines at the poles (Figure 2.14). *

It is worth noting that anywhere in the northern hemisphere the movement of stars immediately above the horizon is always from left to right.

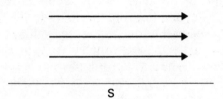

S

Figure 2.14

The only actual circles we can draw on the page which approximate closest to what is seen under the sky in mid-northern latitudes are the paths of those stars which are near the north celestial pole, and do not rise or set in consequence (Figure 2.15). Here our line of sight can be at or near the centre of the stars' turning and a photograph or drawing will represent circles as circles, though without the concave element.

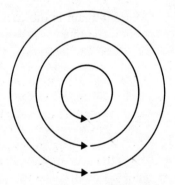

Figure 2.15

The difference between what is actually seen in the sky and what can be represented in the plane is considerable. Not only star paths are changed but also the appearance of star groups. Maps using different projections will give different shapes of constellations. But the celestial sphere is an integral part of the primal phenomena. The professional astronomer's calculations for the positions of stars and planets are made on the model of this sphere – even the sun-centred positions must be derived from it. In the next chapter we shall mention changes in the shapes of constellations which are experienced directly by the observer under the stars, without the intervention of the plane of the page.

To return to movement in relation to the horizon, the swiftest and strongest in this respect is that of stars on or near the celestial equator, a band of them rising either side of the due east point (Figure 2.16) and setting either side of the due west point. The equatorial stars are swiftest or most active

E

Figure 2.16

in that they have furthest to travel, being placed on the largest of the celestial sphere's star circles. A milder and more brief movement occurs to the south, where a passing over of the horizon takes place, distinguished from the thrust of east and west movements. Different again is the circling of the circum-polar stars which remain separate and aloof from the horizon, independent of the zodiac and earth. It is the stars on the celestial equator and within the zodiac which keep the greatest balance between rising above the earth and passing beneath it.

The stars appear to move, then, on a celestial sphere, even though they may be determined by instruments to be at various distances from us. Their distances cannot be determined by the naked eye for the faintest stars are sometimes the closest. However close or distant the star is, our vision virtually projects it onto a huge, but not infinitely distant, sphere (Figures 2.17 and 2.18), for an infinite sphere would be outside the visible realm and would resolve, by projective geometry, into a flat plane. Anyone who has gazed at the sky on a clear night must have felt the amazing 'proximity' of the celestial sphere. In fact the starlit sky at night looks nearer than the blue sky during the day. The clear day-time sky draws one outwards while the clear night sky encloses.

A consequence of the stars appearing to move on a sphere is

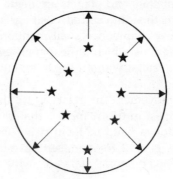

Figure 2.17

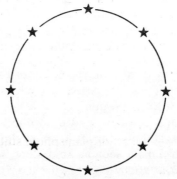

Figure 2.18

that any single star appears to rise or set at the same place on the horizon throughout the year. This is another aspect of the concept of 'fixed' stars, reflected in the word 'firmament'. They maintain a fixed relationship to each other on the sphere and to the horizon (Figure 2.19). Sirius always rises at a certain point on the south-east horizon and sets south-west; Procyon always rises close to due east and sets close to due west; Castor always rises in the north-east and sets in the north-west.

A memorable moment in any early evening is to watch as the blue sky darkens and, suddenly, the first pin-point star is seen, heralding the night. Maybe it is bright Vega, high overhead in early autumn, and one realises it is already visible yet unseen before one perceives it. At first sighting one must look directly at it, yet it becomes surprisingly obvious when looked for again.

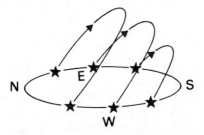

Figure 2.19

It is also felt how much higher it looks when alone, compared with its appearance when its companions are out all around.

At any one time on a clear night in the countryside, from 2,000 to 3,000 stars can be seen, with the average naked eye, above the horizon. Fewer are visible in a city, but surprisingly many still are if one looks carefully; it provides a challenge to identify principal bright stars when their neighbours are hidden by haze or streetlights. Everyone's horizon is different and the student of astronomy should acquaint himself with the stars his sky presents and know them personally at sight like familiar faces. Nothing is better to this end than to discover their identities for oneself from a map or planisphere studied beforehand.

In his autobiography, *Starlight Nights*, the renowned American amateur astronomer Leslie Peltier explained that, for him,

Each star had cost an effort. For each there had been planning, watching, and anticipation. Each one recalled to me a place, a time, a season. Each one was now a personality. The stars, in short, had now become *my* stars.

To conclude on the question of optical impressions: the difference between our normal seeing of terrestrial objects and of celestial objects is that the former is essentially a perspective experience while the latter is not. In 'terrestrial seeing' the vision is drawn in towards vanishing points and lines, whereas in 'celestial seeing' the eye is released from these elements and lives on the sphere. The phenomenon of the stars is therefore essentially two-dimensional and that of earthly objects essentially three-dimensional.

Historically, it took time for the three-dimensional aspect to develop in art and consciousness, as evidenced by two-dimensional representation which pre-dates the Renaissance change to attempting the picturing of three dimensions. Terrestrial seeing not only focuses on the object but on the distance relationships of objects, whereas celestial seeing focuses on objects which stand in space freely and equally, so to speak, without comparison of distances being required – although what appears simply 'behind' or 'in front' in the sky will come into later considerations – for example, when studying eclipses.

* The shapes in Figures 2.11–2.14 are only representative, chosen for the geographical positions indicated, and do not conform to any single method of projection.

Chapter 3
Circling Stars

We are now in a position to describe, visually and qualitatively, certain patterns of stars in the observed sky, and their movements. Firstly we shall turn towards those stars which appear to circle round an imaginary centre point or celestial pole. These are the stars seen when we look northwards from the northern hemisphere of the earth. One group, within the constellation of the Great Bear, moves in a quarter of a circle roughly every six hours (Figure 3.1).

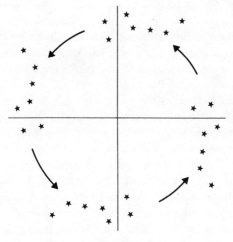

Figure 3.1

It should be noted that the two 'pointer' stars do not align exactly with the celestial pole. Neither are they exactly in line with the so-called Pole Star or Polaris. The pole lies somewhere between this line and Polaris (Figure 3.2). There is no ordinarily

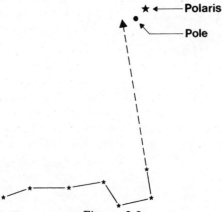

Figure 3.2

visible star exactly at the pole – in fact the pole moves slowly over thousands of years in relation to the stars; this will be dealt with later. The pole approaches closest to Polaris (by about the diameter of the moon) in AD 2100.

The north celestial pole and the Pole Star figure interestingly in history and mythology. In ancient China this motionless region round which all else moved was the throne of a heavenly emperor, the god Shang-ti, who sat behind the emperors on earth who were called 'Sons of Heaven'. In fact when a subject was granted imperial audience, the emperor faced south like the Pole Star and those meeting him faced both the earthly throne and the god at the pole above.

Here we introduce aspects of earlier cosmological pictures as they are part of Man's 'seeing' of the stars on another level – that of their appearance to the imagination which arose out of an earth-centred view. Historically, stories and pictures have always gone hand in hand with observation of the stars as an expression of Man's relation to the universe, so that even very recently groups of stars, or asterisms, within a larger constellation have been given names like the Teapot (in the Archer) or the Keystone (in Hercules). Earlier examples are the Plough or Big Dipper (in the Great Bear) and the Northern Cross (in the Swan).

Such names arose out of constructing a pattern within an area of stars and relating it to some familiar earthly object. This is not, however, how the ancient peoples are likely to have begun naming the actual constellations. Their picture of what stood in the sky in a particular region would have been a pure imagination and a pattern of stars would not have been its outline. For example, it has been suggested that certain early zodiac-division names were related to individual stars, one star representing the whole area – our star Hamal representing, for the Babylonians, the Ram; or early texts denoting the Bull by writing Pleiades or Hyades-Aldebaran. The circular zodiac on the ceiling of an Egyptian temple at Dendera shows zodiacal figures marking areas of sky; later, medieval European painters filled the constellation areas with their artistic impressions of creatures and figures, with stars embedded in them.

On the other hand, joining up stars with lines to form patterns is a natural and useful step to assist recognition in face of the hundreds of points of light presented to the observer, provided he does not trivialise the phenomena by making cartoon figures out of them. Also, it should be realised that the stars were joined up into different patterns by various peoples, past and present.

As to naming the star groups in recent historical times, Erhard Weigel, a seventeenth-century German professor and teacher of Leibniz, proposed a scheme in which the constellations formed the coats of arms of the ruling families of Europe. In the same century two German lawyers, Bayer and Schiller, drew up a 'Christian sky' renaming not only the constellations but the planets too, so that Saturn was called Adam and Jupiter called Moses, etc. More recently, in 1944, the English writer and politician A. P. Herbert published a booklet and map called *A Better Sky* in which the Great Bear became Great Britain with the pointer stars as Shakespeare and Caxton; Cassiopeia became the United States with two of its stars named Tom Jones and Roosevelt, etc.

For observers in early civilisations, however, the stars were presumably contemplated for their positions and qualities of light and movement, and then experienced imaginatively. For instance, six stars in the present Little Bear constellation were known in classical times as the Circlers, Leapers or Dancers round the celestial pole (Figure 3.3). The Egyptians called the circumpolar stars which never set the 'Imperishable Ones' and 'Rowers of the Ship of Ra' – the sun god. They were pictured as swallows which flew back and forth above a heavenly tree, feeding on its immortal fruit and therefore never dying or setting. This area of the sky was also where the pharaohs went

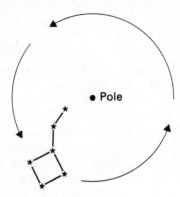

Figure 3.3

after death. Other Egyptian writings give the name of the Jackal to what we call the Little Bear and connect it and the Great Bear with death in that they were conspirators in the killing of Osiris, the Pole Star representing his coffin.

In India a story tells of how the Pole Star is where Prince Dhruva sits meditating so intensely that the whole universe revolves about him.

We can now turn to the wider view of circumpolar stars seen from roughly latitude 52 degrees on earth. We can simplify their relationships by means of circles imagined on the sphere of the observer's sky. One set of circles, concentric to the pole, is an indication of declination (angular distance) between the celestial equator and the pole. Another set of circles passes through the pole and measures the distance the sky turns round the pole in hours, and are called hour circles; each makes one complete rotation about the pole in a sidereal day. The length of a sidereal day is the time it takes a star to complete a circuit round the pole, which is just over 23 hours 56 minutes. This is about four minutes shorter than an average solar day, the latter being used for our clock time.

Both types of circle put together make up a system of co-ordinates to establish the positions of stars in relation to each other and to the observer's sky (Figure 3.4). This system we can use in particular to plot the positions of stars in the northern sky. Projecting this region onto the plane of the page, we produce an elementary map (Figure 3.5). The circles of decli-

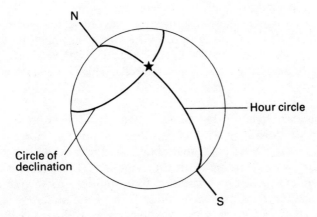

Figure 3.4

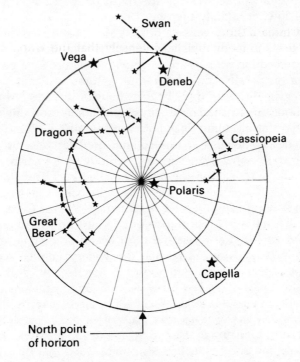

Figure 3.5

nation parallel to the celestial equator remain circles in appearance, while the twenty-four hour circles through the pole appear as straight lines, being seen edge on.

A part of the celestial sphere curving from the horizon to overhead and to left and right of the viewer has been flattened onto a plane. The 'straight' lines in fact curve concavely through the pole like the spokes of an umbrella. This is the sky looking towards the north mapped for 7 p.m. on 1 October at a geographical latitude of 52 degrees north.

The star Deneb in the constellation of the Swan is near the overhead position. The Great Bear is moving downwards towards its lowest point just above the horizon, where it will appear around midnight on the same day, 1 October. As the star sphere turns one revolution four minutes quicker than the sun does in a day, as mentioned before, the sphere will slip round further anti-clockwise each solar day and there will be a time in the year when, for instance, the star Vega in the constellation Lyra will be down at the north point of the horizon, or just below, at midnight (end of December).

The stars of the Great Bear or Ursa Major have been connected with this animal by widely separated peoples in ancient times. This perception of the same imaginative picture for a particular region of the sky among differing cultures is common in the naming of constellations. The early Babylonians saw Ursa Major as a bear, as did the North American Indians, and one notices that the star grouping does not even resemble such a creature. Aristotle explained it by saying that the bear was the animal which could inhabit the cold and solitude of the north.

On the other side of the pole from the Great Bear stands (or rather sits) Cassiopeia, or the Lady in the Chair as the constellation was called by the Greek-influenced Arabs. Greek mythology has it that she was the queen of King Cepheus of Ethiopia. She boasted that she was the most beautiful woman on earth or in heaven, fairer even than the water nymphs. Neptune became angry at this and took his revenge by creating the monster Cetus to plague Ethiopia, resulting in the chaining to a rock of Cassiopeia's daughter, Andromeda, later rescued by Perseus. Cassiopeia was given a seat in the heavens which, however, humiliatingly turned upside down as it swung round the pole day and night. It is also understood that this was the constellation which the Egyptians called the Leg or Thigh which, along with what we now call the Great and Little Bear constellations, was a conspirator against Osiris.

Passing between the two Bears is the ancient constellation of the Dragon which twines part of its length round the pole of the zodiac, to be discussed later. Representing the zodiacal stars by signs for a moment (Figure 3.6) we can imagine that they

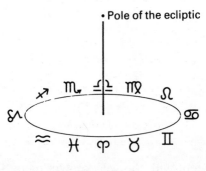

Figure 3.6

possess a celestial pole which, however, is not the same as the north pole (of the celestial equator) already described. Leaving technicalities aside, this zodiac (ecliptic) pole is an important point on the celestial sphere, and the Dragon curls round it. There are many legends about this Dragon. One from Greece describes it as the guardian of the stars which are golden apples hanging from the Pole Tree in the Garden of Darkness. At another time the Greeks described it as having been flung in battle against the star sphere by Athene and caught on the axis of heaven. In Egypt the Dragon was Typhon – or one of Typhon's variants, the Hippopotamus or Crocodile.

The star Vega at the top of Figure 3.5 stands in the constellation of Lyra or Harp, which instrument Orpheus used to charm Pluto, king of the Underworld, to gain the release of his captive bride. This constellation has also been described as one of the birds opposed to Hercules, while the Arabs called it the Swooping Eagle and the star Vega the Falling Vulture. Vega is the brightest star in the northern half of the celestial sphere and has a bluish tint.

Such, then, is a description of some movements and appearances in certain parts of the observer's sky. One last effect of interest should be mentioned – the changing shapes and sizes of constellations depending upon which part of the sky they are seen (Figure 3.7). At first this may seem strange or unlikely,

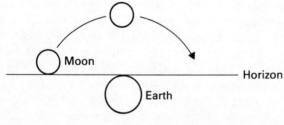

Figure 3.7

but it is consistent with the well-known 'moon illusion' where the full moon when seen close to the horizon appears larger than when it is high above. Photography reveals the moon in both positions to be the same size – and technically, in fact, the moon is almost an earth radius nearer when high above the horizon than when on it (Figure 3.8). Many reasons have been

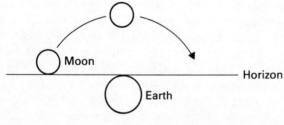

Figure 3.8

suggested for the 'illusion' down the ages. The astronomer Ptolemy (second century AD) proposed that any object seen through filled space, such as the moon seen across terrain at the horizon, is perceived as being more distant than an object just as far away but seen through empty space, such as the moon at the zenith. If the images of these objects in the eye are in fact of equal size, the one that appears farther away will seem larger.*

Another theory was proposed about forty years ago after experiments at Harvard University. This postulated that the apparent size of objects depends on the direction of vision – e.g. turning the eyes upwards diminishes the size of objects. This, however, was discounted twenty years later by further research at Yeshiva University which established by careful tests that from the lying position while looking 'straight ahead' towards the zenith, the illusion is still present – the zenith moon still looks smaller. But the illusion disappeared when the horizon was obscured when viewing the moon 'straight ahead' horizontally, and it also disappeared when an artificial horizon was added to the zenith moon when viewed 'straight ahead' with the subject lying on his back.

This showed that apparent horizon distance was connected with the effect, substantiating Ptolemy's claim. This should not be confused with another theory which says that the horizon moon appears larger because it is seen along with distant objects like houses, trees, etc., and seems large by comparison. This cannot be a satisfactory explanation as the moon still seems larger when low over the sea or over a flat, featureless horizon. Experiments show that what is important is the apparent distance of the horizon – the more distant it is felt to be, the greater the moon's apparent size. This impression of varying distance also takes effect, interestingly enough, in a planetarium where the moon on the artificial horizon looks larger than the one overhead. So the effect has to do with the human being's inner experience of space.

This 'moon illusion' episode is of interest in the light of the distinction made at the end of Chapter 2 between 'terrestrial seeing' and 'celestial seeing'. It harmonises with Ptolemy's solution to say that looking across terrain invokes a perspective impression and this enlarges the apparent size of the moon as perspective carries the sight into distance. This impression is missing on the two-dimensional celestial sphere.

In fact, when the horizon moon is looked at with one eye only, then the illusion of largeness disappears. The three-dimensional, lateral view with both eyes has been changed to two dimensions through one eye, and as the vertical view above is two-dimensional on the sphere, both moons look the same size.

The horizon moon also looks smaller when seen on a mirror (a two-dimensional surface), through a tube, or through a pin-hole in a piece of card.

It is also confirmed that people 'see' the horizon moon as being closer – presumably because it appears larger. But their subconscious picture must be that the horizon is further away than the overhead position on the celestial sphere. This coincides with Ptolemy's view and suggests that our subconscious picture of the sky is of a dome flattened at the top (or half of an oblate spheroid). Experiments support this, dating back to investigations by the English mathematician Robert Smith in 1738. Most people, when asked to raise an arm along a line half-way between horizon and zenith, point in a direction well below 45 degrees from the horizon. A section through this flattened dome is not unlike the arc of vision in Figure 2.2 and also resembles the shape of the top of the human head. The effect the dome has on the apparent size of the moon at different altitudes is illustrated in Figure 3.9 where the angular diameter of the moon is kept the same for the observer and the 'more distant' moon on the horizon looks larger.

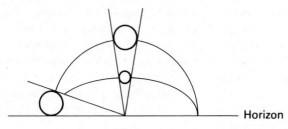

Figure 3.9

An effect of this sort also applies to the apparent sizes and shapes of star constellations. A good example is the Great Bear which appears to grow smaller as it turns upwards from just above the horizon to being high above the pole. Something similar to this effect is produced by the stereographic method of projection used in some star maps which give the effect by chance rather than design because, as previously mentioned, no plane map can reproduce the positional relationships occur-ring on a regular sphere.

To return to the question of circling movement – despite our modern technical knowledge of the motions of the solar system, it is important to appreciate that a rotating earth creates the effect of a moving, spherical space round itself. This effect should not be dismissed. It is an integral part of the observed sky and encompasses and unites the other movements of sun, moon, planets, etc. It is faster than other apparent movements and provides the element within which they swim, with or against its tide. It has its simplest and purest expression in the circling of the northern stars round the throne of the celestial pole.

Copernicus himself, with a qualitative concept of the realm of stars, begins Chapter 1 of his major work on astronomy, *On the Revolutions of the Heavenly Spheres*, with the words:

> First we must remark that the universe is globe-shaped, either because that is the most perfect shape of all, needing no joint, an integral whole; or because that is the most capacious of shapes, which is most fitting because it is to contain and preserve all things; or because the most finished parts of the universe, I mean the Sun, Moon and stars, are observed to have that shape, or because everything tends to take on this shape, which is evident in drops of water and other liquid bodies, when they take on their natural shape. There should therefore be no doubt that this shape is assigned to the heavenly bodies.

Even quantitatively, the modern astronomer visualises a sphere to position the stars upon before deriving everything else from it. For the stars, in the first instance, behave as if they were on a sphere and according to its laws. A celestial globe remains the best star map.

---

* This wording of Ptolemy's opinion is taken from *The Moon Illusion* by Lloyd Kaufman and Irvin Rock.

Chapter 4
Stars Which Rise and Set

The horizon is a dynamic event-threshold where not only the sizes of things seem to increase, as discussed in the last chapter, but where the drama of appearances and disappearances takes place.

To observe a celestial object rising or setting is to form a relationship with it which is quite different in quality to that formed when the object is high above, in the same way that when we meet someone or part from them our experience is not the same as when we relate to them between those times. Also, when an object is near the horizon it meets us on the level of our normal horizontal seeing and the relationship is direct and immediate. When we have to turn our gaze upwards against gravity, the relationship changes – the star or planet enters its own celestial realm and presides there.

Rising and setting are like birth and death and the accompanying greeting and farewell. At the earth's equator all stars move through this birth and death, while at the poles there is an eternal circling.

The Babylonians divided the celestial sphere into three parts named after their gods Enlil, Anu and Ea (Figure 4.1). The

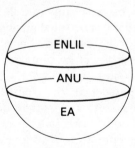

Figure 4.1

upper part of the sphere from the north pole to about four-fifths of the way down towards the equator was apportioned to the air-god Enlil who had separated heaven from earth; the middle band of the sphere on either side of the equator was apportioned to the sky-god Anu; and the lower part down to the south pole was apportioned to Ea, god of earth and water. These were the three celestial 'ways'. The sun in its yearly path then spent three months, or one season, in a particular heavenly region (Figure 4.2).

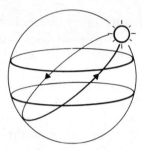

Figure 4.2

In the last chapter we dealt with part of the Babylonian region of Enlil in depicting the circumpolar stars seen from 52 degrees north latitude. The whole region of Enlil would be encompassed by the circumpolar stars at about latitude 73 degrees north, as travelling north raises the pole above the horizon and increases that part of the sphere which turns in circumpolar fashion without rising or setting. Now we shall look at stars seen from mid-northern latitudes which rise and set, which include the middle third of the celestial sphere or the realm of Anu.

Running along the centre of this middle band is the celestial equator, with the zodiacal stars set at an angle to it. The zodiacal stars are those which lie close to the sun's annual path or ecliptic. The ecliptic is therefore tilted to the path of its own motion on the sphere in a day (Figure 4.3). All daily motions on the sphere are 'parallel' to the celestial equator. The best demonstration of the resulting twisting motion of the ecliptic in the course of twenty-four hours is to use a celestial globe, but some indication is given with four of its positions as shown in Figure 4.4.

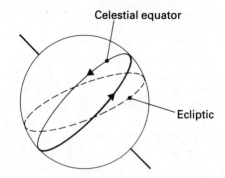

Figure 4.3

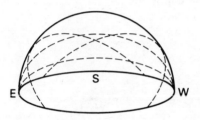

Figure 4.4

Various relationships then arise between the ecliptic and positions on the sphere. One is that which is formed with the horizon, looking south, in the course of twenty-four hours. In periods of six hours the ecliptic (short pieces of it represented by straight lines) appears as in Figure 4.5.

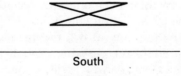

South

Figure 4.5

In twenty-four hours the ecliptic weaves round the horizon and sky, its intersection with the eastern horizon, for instance, occurring to the north-east, east, south-east, and back again. In periods of six hours the relationship of ecliptic to eastern horizon and the equator (broken line) is shown in Figure 4.6,

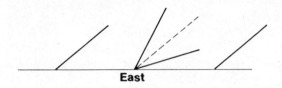

Figure 4.6

and its relationship to the western horizon in Figure 4.7. The ecliptic varies between curving at a particular moment from north-east to south-west, east to west, south-east to north-west, and east to west again; its arc at one due east-west position curves above the celestial equator and at the other east-west position curves below it.

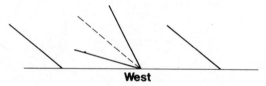

Figure 4.7

Picturing these relationships is essential to understanding the appearances of sun, moon and planets in their risings, settings, and daily or nightly travels. This movement of the ecliptic also expresses, of course, the movement of the zodiacal stars which lie near it. This allows us to characterise the various parts of the zodiac in terms of movement with regard to the horizon. In Figure 4.8, the constellation which lies along that part of the ecliptic marked by the line (a) will rise parallel to the celestial equator (broken line) and pass high overhead. The constellations along the lines (b) and (c) will rise close to the celestial equator and trace out its curve. The constellation along line (d) will pass low over the southern horizon and below the celestial equator.

Stars at line (a) will pass overhead seen from latitude 23½ degrees north on the earth, and this parallel of latitude was called the Tropic of Cancer (the Latin term for Crab) because of this. In earlier times, when the sun stood in the star constellation of the Crab it was overhead at noon at geographical latitude 23½ degrees north. But the stars move over hundreds and thousands of years in relation to our co-ordinates and today the stars of the Twins pass above the Tropic of Cancer.

So the constellation of the Twins is angled parallel to the celestial equator (line (a) in Figure 4.8) as is the constellation

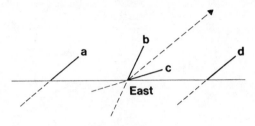

Figure 4.8

along line (d). The difference is that the constellation along line (d) passes low over the southern horizon and in ancient times this led to the naming of the Tropic of Capricorn. Nowadays the stars of the Archer pass overhead at geographical latitude 23½ degrees south.

It will be noticed that while lines (a) and (d) are parallel to the celestial equator and each makes the same angle to the horizon, lines (b) and (c) are not parallel to the equator and each is differently angled to the horizon. These factors determine how fast or slow a constellation will take to rise. The result is that the fastest- and slowest-rising constellations are those which lie near the crossing points of ecliptic and equator (lines (c) and (b)) – i.e. the Waterman and Fishes are fastest, the Lion and Virgin slowest. The other constellations have intermediate speeds of rising (Figure 4.9).

Again, it is best to follow these movements on a celestial globe or planisphere, seeing how the angle of the zodiac to the horizon and equator changes in its rising over twenty-four hours (Figure 4.10). Copernicus tabulated these angles to the horizon in his description of phenomena in *On the Revolutions of the Heavenly Spheres*, which angles express the rotation of the earth about an axis tilted to the ecliptic.

The varying speeds of rising form a qualitative relationship between zodiac and earth with regard to objects' 'arrival' above the eastern horizon. Seen imaginatively, one could picture

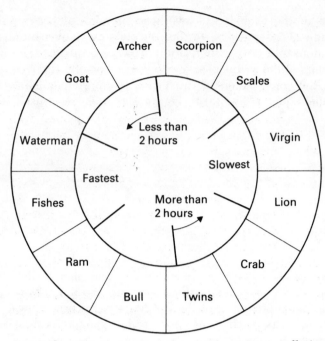

Figure 4.9 Speeds of rising of the visible star constellations of the zodiac

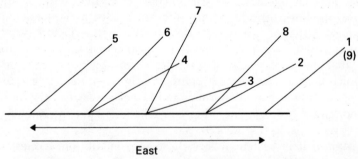

Figure 4.10

speed in terms of colour, with the fast-rising movement tending to the red side of the spectrum and the slow risings tending towards the blue side. One could imagine a sweep of colour round the zodiac expressing the dynamics of the rising constellations. The spring equinox point which marks the beginning of the 'sign' of Aries (not the star constellation of the Ram – see Chapter 5) would stand in the fastest-rising part and correspond to red, and the opposite autumn equinox point would correspond to violet. If the ordinary spectrum colours (rainbow) were spread between these two signs, along the path of the sun, then the summer solstice, for example, would correspond to green. The winter portion of the zodiac could then take on another series of prismatic colours, this time arising out of a dark background (see Goethe's Theory of Colours) with the winter solstice representing peach-blossom or ruby-magenta. This would mean a circle of twelve colours displaying a full range of prismatic hues arising out of light and dark, as do day and night and the seasons, and present a picture of swiftness of zodiacal rising – with dynamic red the fastest, the darkest colours the slowest, and the intermediate colours representing the intermediate speeds. The zodiacal areas between these four cardinal points would show the transition from, say, fast to slow or light to dark, illustrating their character in relation to the eastern horizon.*

One can also differentiate between constellations which cling to the earth in their slow rising and those which readily lift upwards in the east. The stars of the Lion and Virgin take almost three hours to rise, while those of the Waterman and Fishes take less than one hour. In early times, that section of the zodiac which was rising was considered a most important aspect of a horoscope – in fact this rising section was known as the 'horoscope' and the word was only later applied to the whole chart. These remarks are in place as old astronomy and astrology were inextricably bound up with each other. The fast-rising parts of the zodiac were referred to as being of 'short ascension' and the slow-rising parts were of 'long ascension'.

Interestingly enough, the sequence of rising phenomena reverses for setting. For instance, that part of the zodiac which inclines steeply when appearing above the eastern horizon, has a shallow incline when disappearing below the western horizon. Constellations midway between the extremes of angle rise and set at the same inclination.

Another aspect in forming a qualitative picture of the zodiacal stars is to consider their compass positions of rising and setting on the horizon, and their high or low passage across the sky. In mid-northern or mid-southern latitudes the zodiacal stars describe a living, weaving movement in relation to the earth, while at the geographical equator and poles the extremes take over; at the equator there are continually dynamic risings and settings with the zodiac upright against the horizon and taking a high passage above, while at the poles there is no rising or setting of the zodiac and half of it circles low round the horizon, maintaining the same altitude.

Whenever the zodiac or its constellations have been mentioned so far, the visible stars bearing those names are thereby referred to, with equal divisions given to each constellation. There are other zodiacs, including the tropical zodiac 'signs', which will be discussed in the chapter on the sun (Chapter 5).

We now turn to a few individual stars and constellations. Four stars standing in separate constellations mark out approximate quarters of the sky. They are Aldebaran, Regulus, Antares and Fomalhaut. The first three lie within the traditional zodiac but Fomalhaut does not. Copernicus gave Regulus its name (derived from its earlier name Rex) and it was chief of the Four Royal Stars of the ancient Persians, their Four Guardians of Heaven. The French astronomer Flammarion identified the other three royal stars as Aldebaran, Antares and Fomalhaut.

Greeks, Romans and Arabs called Regulus the Lion's Heart, and indeed it lies in the constellation of the Lion. The Chinese called Regulus the Great Star in Heen Yuen, a constellation named after the imperial family. Its colour has been described expressively as 'flushed white and ultramarine' in the days before the quantitative unsubtlety of colour photography and index numbers, which can never capture the nuances of what the naked eye sees. The Egyptian King Necepsos and his philosopher Petosiris taught that the sun was created in the stars of the Lion. All major ancient cultures from Persia to Greece referred to this part of the zodiac as the Lion, and connected it with the sun. In 3000 BC the sun at midsummer stood among the stars of the Lion. Regulus is almost directly in the path of the sun which only just misses covering it every year.

Aldebaran and Antares are unique in that as bright stars (both of reddish colour) they stand almost exactly opposite each other in the zodiac. Aldebaran, described as pale rose to the naked eye, stands in the middle of the Bull and the fiery-red Antares stands in the Scorpion. This latter constellation has been known as the birthplace of Mars, and Antares itself is thought to have derived its name from the Greek for 'rival of Mars' or 'similar to Mars'. Alchemists said that only when the sun was in this area could iron be transmuted into gold.

The name Aldebaran is from the Arabic *Al Debaran*, the Follower – i.e. of the Pleiades. The nearby Pleiades is a beautiful, hazy cluster of stars (six normally visible to the naked eye) and they seem to be among the first stars mentioned in astronomical literature (in the Chinese annals of 2357 BC) and have featured prominently in the history and literature of many cultures ever since.

The importance of Aldebaran and Antares lying almost exactly opposite each other in the zodiac is that the twelve zodiacal constellations can be positioned and measured from them round the circle. Robert Powell and Peter Treadgold, in their book *The Sidereal Zodiac*, have researched this relationship and concluded that the Babylonians based their star constellation zodiac (as opposed to the later tropical zodiac) on these two stars and they place Aldebaran (as at the year AD 1950) exactly in the middle of the Bull, measuring the star constellations in equal areas of 30 degrees from there. This results in a sidereal zodiac as shown in Figure 4.11 which the present author uses when referring to a sidereal zodiac of equal divisions. A. Pannekoek, in his classic *A History of Astronomy*, also states that the Chaldeans developed a zodiac of equal divisions based on star positions. A sidereal zodiac of unequal divisions which dates back to the star charting of Ptolemy is normally used by the modern astronomer.

The fourth Royal Star of the Persians, understood to be Fomalhaut, is of reddish colour** and stands in the constellation of the Southern Fish. Its name, from the Arabic *Fum al Hūt*, means Fish's Mouth. In early legend the Southern Fish was the parent of the zodiacal constellation of the Fishes. It lies beneath the constellation of the Waterman and has been depicted as drinking water flowing from the latter's jar. In mid-northern latitudes Fomalhaut passes very low over the southern horizon but its brightness is enhanced by the absence of other nearby bright stars.

The ancient Egyptians observed the rising of the zodiac in stages of ten days each (decans) with thirty-six decan stars marking out the year. As a decan star makes its first appearance of the year above the horizon at dawn on a particular date, a division of the adjacent zodiac is indicated (Figure 4.12). The

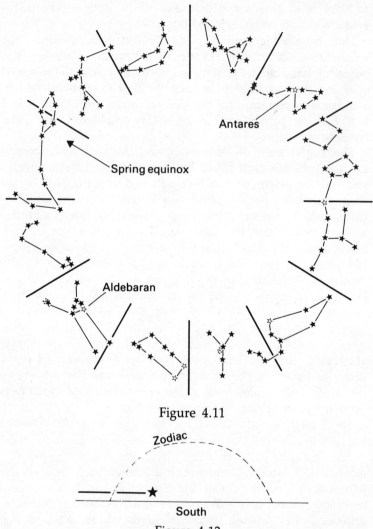

Antares

Spring equinox

Aldebaran

Figure 4.11

Zodiac

South

Figure 4.12

decan stars were positioned in an area of the sky to the south
of the zodiac. Sirius, the brightest of all stars in the sky, marked
the thirty-sixth decan and a position on the zodiac beside the
stars of the Lion. Decan stars like Sirius leave the night sky
for more than two months between their evening setting and
morning rising with the sun. An Egyptian papyrus describes
the decan star as remaining 'in the underworld, in the house

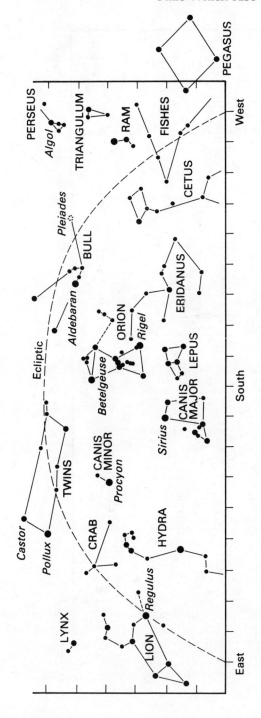

Figure 4.13 Midwinter midnight sky for latitude 52 degrees north

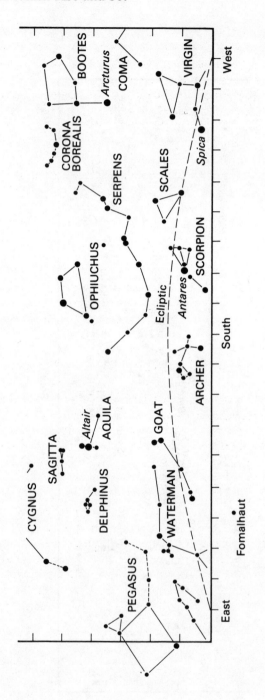

Figure 4.14 Midsummer midnight sky for latitude 52 degrees north

of Geb' (the earth god) for seventy days where 'it purifies itself and rises on the horizon like Sothis' (Sirius). One is reminded of a description in Cairo Museum of the embalmment of Tutankhamen in which his body was prepared for burial by the priests for seventy days, during which time hymns were chanted and prayers and spells recited for the soul of the dead.

We can now depict on a map the zodiacal stars and surrounding constellations as seen above the horizon (Figures 4.13 and 4.14). The northern stars, as shown in Figure 3.5, turn free of the horizon with their co-ordinates of latitudes and hour circles turning with them. But the stars which rise and set relate themselves strongly to the horizon, so our co-ordinates this time are fixed in relation to the earth, and the stars move behind them. These co-ordinates are called azimuths (compass points along the horizon) and altitudes (heights above the horizon).

Two positions have been chosen for the zodiac in the course of the year – the stars seen at midnight in midwinter and those seen at midnight in midsummer. This allows us to view both halves of the zodiac and see two of its extreme positions with regard to the horizon. The zodiac assumes these positions once every day, and the positions have the sun at an opposite side of the sky (beneath the northern horizon) only once a year.

Sirius can now be seen in Figure 4.13 near the constellation of Orion, the latter associated with Osiris by the Egyptians. Sirius itself they connected with Isis. A prominent star in the sky as depicted in Figure 4.14 is pale-yellow Arcturus which, in 1933, demonstrated an influence from the stars when its light was transmitted via a 40-inch telescope to operate a switch which turned on the lights of Chicago's Century of Progress Exposition.

One geometrical aspect of the relationship between the zodiac, or ecliptic, and the horizon is that, seen from the earth's equator, the mid-point of the ecliptic or its highest point describes a lemniscate (figure of eight) on the celestial sphere in a day (Figure 4.15). At the poles this form smoothes out into a straight line (circle) above the horizon, with intermediate forms at intermediate latitudes. Other connections with the lemniscate will be dealt with later; in particular concerning the apparent loops of the planets. The above lemniscate is essentially in connection with the sun's path. For a yearly representation of this form at the equator, one would plot the

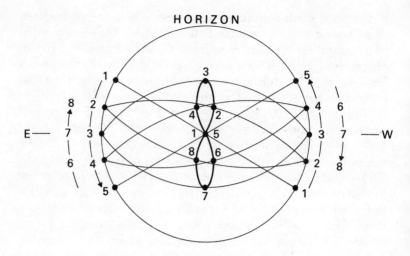

Figure 4.15

mid-point of the ecliptic arc for a particular time of day throughout the year, say midnight.

Astronomers describe the rising and setting of stars, planets, sun and moon in the context of atmospheric refraction or the shift upwards given to celestial objects above the horizon through the deflection of their light by the density of the air. The elimination of this effect, called a correction, is necessary in order to calculate positions accurately on the celestial sphere for navigation, surveying, etc. An object just above the horizon by observation will be below it by calculation, this calculated position being called its 'true' one. One also speaks of a 'technical' horizon as opposed to the visible one. Yet the visible position of a star cannot be calculated exactly due to variations in atmospheric density. Technical and observed positions coincide when an object is overhead and are most separated from each other at or near the horizon. At the zenith the apparent and the ideal are one.

The result of this subtle effect is that rising objects will appear higher than their technical positions, then as the effect wears off with increasing altitude above the atmosphere, they will fall back towards their technical positions. Thus rising and setting objects will linger a little in the vicinity of the horizon. This is built-in to the experience of the earth-based observer and is part of the drama of the moving sky. It also raises the question of 'where' a star really is. But if a tennis player strikes a ball downwards to the opposite side of the court so that it deflects upwards towards his opponent, the opponent cannot say the ball is not travelling towards him just because it was not struck directly towards him in the first place. A refracted star is shining towards us when we see it and 'is' where it is seen. This is its 'true' position. It then slows down its motion from where we stand. Otherwise one falls into the habit of saying that though a star is visible, it has not yet risen or has already set. We should not phase out the dynamic element in observation too readily and should allow, when appropriate, the ideal concept of the celestial sphere to flatten as if it were a roof above our heads and become like an oblate spheroid (Chapter 3);*** and we should allow it to be flexible enough to slow down slightly near the horizon. Visually, due to refraction, there is always more of the star sphere above the horizon than below it, to the extent of about one moon diameter.

Atmospheric effects also result in stars twinkling, caused by variations in atmospheric density at different heights. Sirius, never achieving a very high altitude at mid-northern latitudes, is conspicuous in its flashing out the colours of the spectrum. If a star is high above the horizon it achieves a steady, celestial light. Another effect of atmosphere is that sun, moon and planets take on a yellow or reddish tint due to a darkening of their light.

Disregarding refraction for the moment, the observer's horizon (a plane which is tangent to the surface of the earth)

CELESTIAL SPHERE

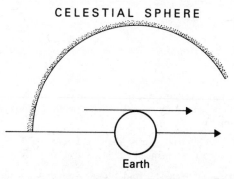

Earth

Figure 4.16

is, as far as the stars are concerned, also an earth-centred position; it has the same star hemisphere arching above it as has a plane through the centre of the earth (Figure 4.16). The reason is that a star is far enough away for any straight lines directed towards it from different parts of the earth to be virtually parallel.

This observer's horizon, as suggested earlier, is a dynamic threshold where size, colour and speed work in a special way. Altogether the rising, passing over and setting of stars is a rich and varied pageant for the perceptive or informed onlooker.

---

* See also the zodiacal colour circle described in chapter XXII of *The Basic Principles of Eurhythmy* by Annemarie Dubach-Donath (London, Rudolf Steiner Publishing Co., 1937).

** To the naked eye it often appears reddish, caused by the atmosphere at its low position in the northern hemisphere. Objectively it is white.

*** The earth, too, is an oblate spheroid. Stones worn down by the sea tend to become a similar shape – i.e. ellipsoids.

Chapter 5
The Sun

The sky and the human eye are spheres with parts in common which act like refracting lenses. The atmosphere round the earth refracts light on the horizon to the extent of about the sun's apparent diameter (roughly half a degree of arc), allowing us to see more than a hemisphere of stars. This also means that we see more of the sun than we would otherwise, and on average it appears above the horizon for a longer time, in the course of a year, than it spends below.

The corneal surface of the eye refracts extra light into the eye and the human being can see, peripherally, at an angle wider than 180 degrees. A small central area at the back of the retina, about 1.5 mm in diameter, is called the yellow spot and possesses yellow pigment. Within this is the fovea centralis, a depression about 0.3 mm in diameter which gives us the most sharply defined vision in daylight and is also the most important region of the retina for colour perception. The diameter of the fovea centralis, when projected through the eye, covers an area on the celestial sphere of about 1 degree. This corresponds to the apparent diameter of two suns or sun plus moon. Eye and celestial vault together form a type of spherical lemniscate with sun and moon on the external hemisphere and the yellow hollow of the fovea centralis on the hemisphere within.

In the course of twenty-four hours the sun moves against the stars a distance, again, of about 1 degree, or twice its diameter. This movement is towards the east, which means that the sun will rise a little later each day (Figure 5.1) – by the amount of time it takes the celestial sphere to move through one degree of arc, which is about four minutes in time. A Babylonian measurement of time was in intervals of four minutes, called uš. This registers the difference betweeen the sidereal day and

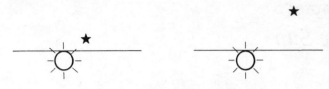

Figure 5.1

the solar day. Again, from sunrise to sunset on an average day (twelve hours) the sun moves its own diameter against the stars, while it moves through an average of about 360 diameters in its diurnal movement from eastern to western horizons.

The rising and setting of the sun means that there is a perpetual dawn and dusk rotating round the earth where light meets darkness (Figure 5.2). Travelling with this shadow edge

Figure 5.2

are the reddened colours of the sky, and the dawn and dusk chorus of birds. This also means that all times of day or night are perpetually somewhere on earth – for instance, at every moment there is a midnight and a midday stretching along the great circle which passes through the centre of the shadow and the centre of the lit part. In fifteen minutes the shadow's edge will have passed from Amsterdam to London while it will take about another four hours to cross the North Atlantic.

Because the sun does not keep the same height above our horizon from day to day, this brings about the seasons. In relation to the earth, the sun spirals up and down round the axis of the poles in each year (Figure 5.3). When the sun is half-way up or down the spiral the shadow passes through the poles and night and day are equal everywhere (Figure 5.4), hence the

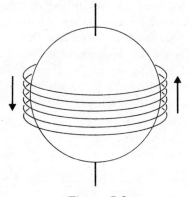

Figure 5.3

Figure 5.4

term 'equinox'. When it is at the most northerly point of its movement it is summer in the northern hemisphere (which is most lit) and when it is at the most southerly point it is summer in the southern hemisphere (Figure 5.5). Sun, moon and planets spiral round the earth in a similar way that consecutive leaves spiral (as Goethe realised) in their growth round and up the stalk of a plant.

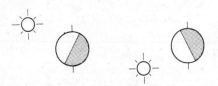

Figure 5.5

In addition, the sun is closer to the earth when at its most southerly point, and a measurement of its diameter in winter in the northern hemisphere will show that it is larger than in the summer. It is not the closeness of the sun to the earth which causes the seasons, but the angle at which the sun's warmth meets the earth's surface. Also, the more atmosphere which the warmth has to pass through, the cooler the season (Figure 5.6).

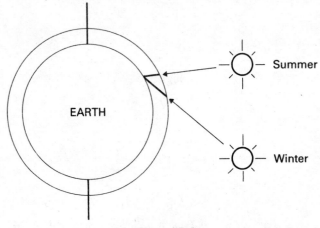

Figure 5.6

Therefore the tilt of the sun's path (ecliptic) to the earth's equator* regulates the intensity and rhythm of our seasons in conjunction with the air's breathing element in how much warmth it allows to pass through. It also means that at any one time, opposite seasons exist on earth. At the equator, day and night alternate in roughly twelve-hour intervals throughout the year, and in fact it is difficult to determine the season astronomically there. The sun seen from the equator is highest in the sky when it stands at the equinoxes and lowest when it stands at the solstices (Figure 5.7). At the poles another extreme is reached in that the year consists of only one day. There, the sun is above the horizon for 189 days and below it for 176 days. The two-week period which the sun spends above the horizon in excess of the time it spends beneath it is caused by inequality of the seasons and refraction. At the poles dawn and dusk last for over seven weeks each. This is when the sun is less than 18 degrees below the horizon (astronomical twilight). Therefore

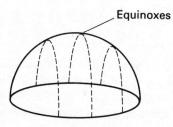

Figure 5.7

dark night as such lasts for about 2 and a half months at the poles, after which there are almost two months of dawn.

The shortest duration of twilight takes place in equatorial regions, where the sun meets the horizon at an angle of 90 degrees. Here it lasts about an hour. Elsewhere, apart from the polar regions, longest twilight takes place at midsummer, the twilight lasting longer the further north one goes. Beyond 48½ degrees north or south, which includes the whole of Britain, twilight at midsummer lasts all night. The shortest duration of twilight at any place depends on its latitude and there are two shortest times in the year – before and after midwinter. This is because twilight depends not only on the path which the sun travels beneath the horizon but also on its speed on the celestial sphere. Objects travel fastest, as explained earlier, on the celestial equator, so the midwinter movement of the sun below the horizon will be relatively slow, thus lengthening twilight. At a geographical latitude of 52 degrees, shortest twilight will last 1 hour 50 minutes in the middle of October and the end of February.

Twilight is a most sensitive part of the day, bringing colour into the sky, songs into the throats of birds, and a delicate atmosphere as nature seems to pause and the human being reflects.

The Egyptians, according to the astronomer Norman Lockyer, said that dawn and twilight were the goddess Isis. Her offspring was Horus, the rising sun, which became Ra at midday and Osiris (husband of Isis) at sunset. The setting was a death, Osiris being killed by the circumpolar constellation god of darkness, Typhon. Horus then rises to avenge the death of his father, Osiris, and kills Typhon. This is a simplification of Lockyer's account, but it indicates a mythology in which the direction in space or the moment in time are themselves deities.

Individual stars were designated as goddesses but when rising just before sunrise they were represented as Isis nursing the young Horus.

Although setting was a death for the sun, a special kind of extinction took place for a star or planet in the ancient world between when it disappeared into the sun's rays at sunset and reappeared on the other side of the sun at sunrise (Figure 5.8).

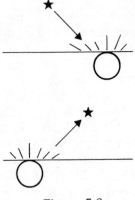

Figure 5.8

This departure from the evening sky once a year and re-entry into the morning sky is called the heliacal setting and rising. The period of invisibility, which varies and lasts for days and weeks, is associated with the Babylonian 'hiding places' of the stars and planets by the writer Cyril Fagan. The Egyptians mourned the heliacal setting of a star and celebrated its rebirth at heliacal rising.

All celestial bodies become invisible to the naked eye when entering far enough into the sun's realm of light, and at midday the light reigns supreme. True midday is when the sun is due south and at its highest above the horizon for that day. Midday means, in the first instance, the middle of the daylight period, but it has not always marked the middle of the twenty-four hour period. At an early stage the Babylonians measured their days from sunset to sunset, divided into twelve periods, each called a 'beru' and lasting about two hours. Later they divided the period from sunset to sunrise into twelve equal parts, and likewise from sunrise to sunset. Therefore in winter the twelfths or 'hours' of darkness were longer than those of daylight, and in summer the daylight 'hours' were longer. Aspects of such a system became reflected in Greek, Roman and European practice, and there are sundials showing these 'unequal' hours throughout the year (e.g. at the Old Royal Greenwich Observatory) called temporal hours, which prevailed for civil purposes in Europe during the Middle Ages. The day was also measured in early Europe from sunrise, or the mean sunrise of 6 a.m., so that the time of Christ's crucifixion, 3 p.m., was written as 'the ninth hour'. Apart from civil time, the astronomer down the ages has often preferred to use midnight or midday for the starting point of a daily cycle.

Early divisions of the day inclined to follow the visual position of the sun, to move with it throughout the year. Later this 'true' or actual timing for a particular place was changed as swifter communications grew between territories. This brought in 'mean' or average time which a wide area could share, and in England was called 'railway time' and was adopted nationally in 1880. Before then the traveller had to adjust his watch as he travelled east or west, the sun marking noon at Bristol ten minutes after having done so at London. Mechanisation, in the form of clocks, brought a second distortion of the true day by making every day the same length so that circular cogs and wheels could better deal with it. But the apparent sun speeds up and slows down rhythmically in the course of the year so that days are a little longer in winter in the northern hemisphere than they are in summer. This results in mean time and true (apparent) solar time falling out of step during the year. A second reason for this effect is that the sun is not travelling along the celestial equator, which keeps an even path round the earth, but travels along the ecliptic which is at an angle to the equator and does not share its divisions. Mean time, so to speak, places the sun on the celestial equator all year. The disparity between mean and apparent time is called the equation of time. Figure 5.9 illustrates this; the lemniscate is traced out by the tip of the shadow of a vertical stick in the ground at mean middays. At apparent noon during the year the tip of the shadow would move up and down a straight line lying due north from the stick. The lemniscate shows the difference between a sun moving on the circle of the celestial equator (mean) and a sun moving on an ellipse in the ecliptic

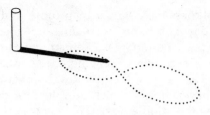

Figure 5.9

(apparent).** Four times a year the two suns show the same time – on about 16 April, 14 June, 1 September and 25 December. In February the mean sun goes ahead of the true sun by 14 minutes 21 seconds; in May it falls behind by 3 minutes 45 seconds; in July it goes ahead by 6 minutes 22 seconds; in November it falls behind by 16 minutes 22 seconds. This lemniscate, in fact, represents a dead, fixed time system in comparison with a living one in which the sun's shadow breathes in the element of time by reaching its noon mark now swifter, now slower in the course of the year – a rhythm within which nature and the human being unconsciously live.

The length or axis of the noon-mark lemniscate is determined by the height the sun stands at the summer and winter solstices. This is also expressed in the position on the horizon at which the sun rises or sets. Figure 5.10 shows (diagrammatically) rising

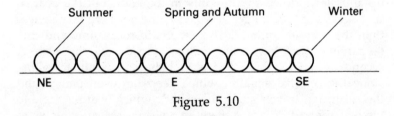

Figure 5.10

positions throughout the year in mid-northern latitudes. The angle along the horizon between the solstices (stations or stopping points) is astonishingly wide, reaching around 80 degrees in southern Britain. It is a surprise to become aware of the swing along the horizon which risings or settings make through the seasons. Likewise, though more obvious, is the sweep upwards and downwards which the noon positions of the sun make in the south (Figure 5.11).

The great Greek astronomer Hipparchus (c. 190 BC to 120

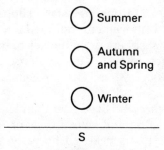

S

Figure 5.11

BC) is thought to have been the first to have systematised a slow movement of the celestial sphere which, over centuries, noticeably changes the sun's regular seasonal positions among the stars. In 129 BC he found the star Spica in Virgo to be 2 degrees nearer the autumn equinox point than was observed by Timocharis either 154 or 166 years earlier (the date is uncertain). At that time Spica was west of the autumn equinox and the difference in position suggested that the celestial sphere, in addition to revolving westwards daily, also slipped back slowly in an eastward direction round the axis of the ecliptic. Therefore the zodiacal stars would pass in turn through the equinox and solstice points (Figure 5.12) or, alternatively, the equinoxes and

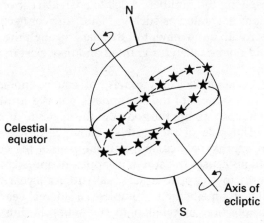

Figure 5.12

solstices would move through the stars. Hipparchus' calculation for the speed of this movement indicated that the time it would take for the celestial sphere or equinoxes to move one complete revolution was (according to later evidence of Ptolemy) either 27,692 years or 29,862 years. The present-day estimate of this cycle is 25,770 years.

Ptolemy, however, who lived later than Hipparchus, was less accurate than the latter and gave a cycle of 36,000 years which is identical to what was believed by early astronomers to be Plato's 'Great Year'. Plato defines the Great Year (in *Timaeus*) as the period after which all the planets and stars repeat their relative positions. Whether he means this to be a period of 36,000 years is not easily deducible from his writings. But this number is said to be mentioned, for example, by the English mathematician and astronomer Sacrobosco (c. 1190 to c. 1255) who is quoted in a presumed version of his *Tractatus de Sphaera**** – 'the ninth circle in a hundred and a few years, according to Ptolemy, completes one degree of its own motion and makes a complete revolution in 36,000 years (which time is commonly called a great year or Platonic Year)'. Although Ptolemy linked this number with the slipping back of the celestial sphere one revolution, it is not certain that a connection with a 'Platonic Year' is valid. Ptolemy's figure for the star movement was quite inaccurate and applying the number 36,000 to a Great Year of Plato is a supposition, not stated as such in Plato. Any connection of a Platonic Year with a movement of the stars through the sun's seasons should not be taken for granted and may exist in name only.

Today the cycle of 25,770 years is, of course, no longer described as a movement of the star sphere, but as a turning of the celestial equator through the zodiac in the opposite direction (westwards). The pole of the earth's equator is described as moving in an approximate circle round the pole of the ecliptic (Figure 5.13). This causes the sun in its apparent annual motion round the earth to complete a seasonal (tropical) year about twenty minutes in time before it completes a sidereal year (i.e., returns to its position in relation to the stars) as illustrated in Figure 5.14. Therefore the sun precesses (precedes) slowly westwards in its seasonal positions in relation to the stars. The two intersections of the celestial equator with the ecliptic at the sun's spring and autumn positions thus move westwards

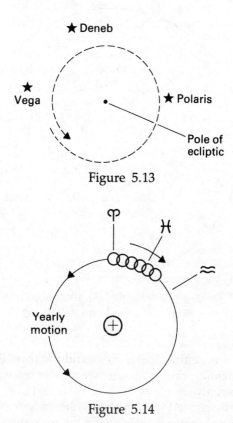

Figure 5.13

Figure 5.14

against the stars and the movement is called the precession of the equinoxes. Figure 5.15 shows a modern estimate of the positions of the spring equinox in the equal zodiac constellation system of the Babylonians.

In Mesopotamia and Greece in the two or three centuries before Christ, the zodiac was divided into twelve equal 'signs'. The Chaldeans measured along the ecliptic from a star position, so that the scale of ecliptic degrees was kept fixed to the stars. The Greeks, however, measured the ecliptic from the spring equinox (or other seasonal positions) so that the ecliptic divisions moved westwards against the stars. If one called the first 30-degree division after the spring equinox the sign of Aries, then this sign must, with the precession of the equinoxes, move through the star constellations of the zodiac. Today, for instance, the sign of Aries stands among the stars of Pisces. But

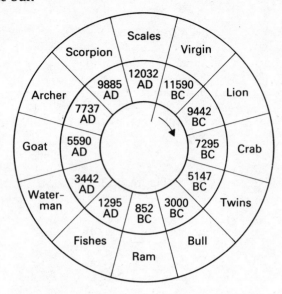

Figure 5.15 Dates for the spring equinox standing at the centre points of equal constellations

in Ptolemy's time the sign of Aries and the stars of Aries more or less coincided, causing controversy later over which zodiac he must have used.

Broadly, there are three zodiacal systems to choose from and Ptolemy probably used all three. Each has its own validity, and grew out of man's evolving relationship with the sky. Firstly, perhaps, there is the contemplation of the stars for the pictures they form in the imagination. Qualities of light, position and movement of individual stars and groups of stars is foremost, with little or no consideration given to measurement on the celestial sphere. Later this develops into painted pictures, of the creatures imagined, on a star map with the figures covering unequal areas of the zodiac. As mentioned already, this was one of Ptolemy's ways of depicting the sky and is the basis for the unequal zodiacal constellations of modern astronomy.

Secondly, there is the equal division of the zodiacal constellations known to have been used by the Babylonians. Measurement and calculation here become as important as the constellation figures.

Thirdly, there is the Greek division of the ecliptic into twelve equal parts measured from the sun's seasonal points (e.g. the spring equinox). This is a moving zodiac, its 'signs' shifting against the stars with precession (Figure 5.16).

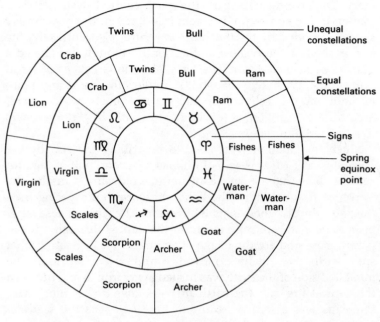

Figure 5.16

The latter tropical zodiac is based on the sun and was introduced later in history when the development of individual consciousness was greater. It was used by the astrologer who wanted to calculate the horoscopes of individuals. The sun and the personal element seemed to go together. The personal element was the one which moved independently of the stars. This is a qualitative assessment of the historical validity of the tropical zodiac besides its facilitating geometrical simplicity, etc. The heavens and human destiny were not considered apart in early times. Essentially, the tropical zodiac emphasised the sun whereas the sidereal zodiac was, of course, oriented to the fixed stars which maintained a constant relationship to the earth. In the Babylonian culture in which this zodiac was used, predictions concerned nature, the land and social and political questions. The situation was less personal and more objective, more to do with the life forces of nature and man. Aristotle called the zodiac the 'life-bringer' and each zodiac would have been considered to carry this out in its own way.

The sun is both planet and star, depending on which planetary system one uses – in the earth-centred system it is a planet, in the sun-centred system it is a star in the present-day sense. In both systems the sun is a centre. Geocentrically, the sun is chief of the planets and anciently, though designated a planet, was called a star. The term 'star', in Greek *aster* (as in our words 'asteroid' and 'astral'), applied to almost any point of light on the celestial sphere, even comets which were, according to Apollonius (c. 262 BC to c. 190 BC) 'proper stars like the sun and moon'.

Much imagination was put into describing the sun by the ancient Greeks. It was said of Philolaus (second half of the fifth century BC) that he spoke of a threefold sun – firstly, the universal fire; secondly, the light and warmth emanating from this; and thirdly, the beams which reach the earth. These latter beams, he said, are reflected from the sun as through a transparent glass and are the image of an image. He also taught that sun, earth and planets moved round a central fire. The inhabited side of the earth was turned away from it and between the central fire and the earth there revolved a counter-earth. Philolaus was a Pythagorean and here one sees the seeds of Copernicanism.

At night we stand in the shadow of the earth. If we look carefully in clear conditions, the shadow can be seen at sunset curving in the atmosphere above the western horizon. Later in the evening it remains invisible above the horizon, stretching towards the ecliptic. Occasionally it becomes visible when the illuminated moon enters it in an eclipse (origin of the word 'ecliptic'). It also takes an effort to realise that the dark night sky is filled with sunlight. It simply needs the moon or a planet or a satellite to enter the sky for us to see it lit up in the sunlight. But light itself is invisible. We only see illuminated objects, not light. The human being can see, as he does a faint star, a candle flame from a distance of 27 kilometres, but what he 'perceives' is illuminated gas in the flame. Sunlight can be understood in the same way.

Light is a mystery and a distinction has to be made between manifest and unmanifest light. We claim to detect light, to quantify it, to measure its velocity, and yet light itself may not in fact be accessible to any of these things. We perceive its effects and incorporate these effects into our experience of the physical world. Somewhere there is an illusion involved, and the ideas of Philolaus may prove to have more in them than meets the eye. Subtle questions on the nature of light are raised by Ernst Lehrs in his book *Man or Matter* and, in this connection, aspects of Einstein's Theory of Relativity are criticised in *Science at the Crossroads* by Herbert Dingle and *The Einstein Myth and the Ives Papers – A Counter-revolution in Physics* edited by Dean Turner and Richard Hazelett. As yet such publications are not taken very seriously by orthodox scientists.

An early enthusiast on the nature and mystery of the sun was Julian the Apostate, Roman emperor from AD 361 until his death in battle against the Persians two years later. Connected with the Mysteries of Mithras, he propounded a threefold sun – firstly, the archtypal offspring of the Good in the realm of Ideas; secondly, Helios, ruler of the intellectual gods; and thirdly, the visible sun.

Julian wrote a 'Hymn to King Helios' in which he spoke of the sun's beams culminating in the Fifth Substance or ether. '[I]t is hard,' he said, 'as I well know, merely to comprehend how great is the Invisible, if one judge by his visible self.' Julian's writing has behind it an intense inner experience of the stars from his early years and, whatever his opinions, he reveals the best prerequisite for a study of astronomy – a sense of wonder.

On the first page of his 'Hymn to King Helios' he says:

> [F]rom my childhood an extraordinary longing for the rays of the god penetrated deep into my soul; and from my earliest years my mind was so completely swayed by the light that illumines the heavens that not only did I desire to gaze intently at the sun, but whenever I walked abroad in the night season, when the firmament was clear and cloudless, I abandoned all else without exception and gave myself up to the beauties of the heavens; nor did I understand what anyone might say to me, nor heed what I was doing myself. I was considered to be over-curious about these matters and to pay too much attention to them, and people went so

far as to regard me as an astrologer when my beard had
only just begun to grow . . . But let what I have said
bear witness to this fact, that the heavenly light shone
all about me, and that it roused and urged me on to its
contemplation, so that even then I recognised of myself
that the movement of the moon was in the opposite
direction to the universe, though as yet I had met no one
of those who are wise in these matters.

Here the inner human experience and the phenomena are
united. Many a scientist will have begun his interest in nature
with an enthusiasm and feeling such as Julian the Apostate's,
but found it difficult to maintain it in an age of external quanti-
fication. Amateur astronomers are an example where this initial
feeling is often kept particularly alive through years of unsophisticated observation in all weathers under the starry sky.

The ancient Babylonians described, in their texts, the daily
movement of the sun as 'zi sha Shamash' or 'life of the sun god'.
Heraclitus of Ephesus (c. 500 BC) is reported to have said that
the sun is kindled at its rising and extinguished at its setting,
that 'the sun is new every day'. Such sentiments are antiquated
to the modern mind, but if we are to penetrate the meaning of
celestial phenomena for the human being, then such sentiments
must be replaced with others equally imaginative and as new-
kindled every day as Heraclitus' sun.

* This angle, the obliquity of the ecliptic, varies and is at present about 23.44
degrees. It is decreasing, reaching a limit of about 22.58 degrees in roughly
12,500 AD. In about 7,500 BC it was at a higher limit of around 24.25 degrees.

** The eccentricity (deviation from a circle) of the earth's elliptical orbit
oscillates and is at present decreasing over thousands of years.

*** Scholars disagree over this source.

Chapter 6
The Moon

Man's view of the universe changes with his development and
there is no better demonstration of this than in the science of
astronomy. Man's view of the moon has variously seen it as –
beyond the stars; closer than the stars; a male deity; a female
deity; as parent of the sun; as offspring of the sun; having its
own light; having only reflected light; being a globe with one
half lit and the other half dark; not being a solid body at all;
being a mineral body once part of the earth; being a mineral
body with origins outside the earth; being a planet of the sun,
describing a curve round it which is never convex but flattens
twelve times (modern view); and so on.

The further back one goes in history the more important the
moon appears to become in terms of influence. The Roman
Emperor Augustus had a silver coin issued during his reign
which bore the sign of Capricorn – i.e., where the moon stood
at his birth on 23 September 63 BC. The oldest document
surviving of *scientific* astronomy from Babylonia, dated 523 BC,
lists information on the first and last moon crescents of the
months and on phenomena around full moon. Such details had
religious and predictive importance, as indicated by an earlier
Assyrian text:

> When the Moon reaches the Sun and with it fades out
> of sight . . . there will be truth in the land and the son
> will speak the truth with the father. On the 14th the god
> was seen with the god . . . when the Moon and Sun
> are seen with one another on the 14th, there will be
> silence, the land will be satisfied; the gods intend Akkad
> for happiness

or 'when the Moon does not wait for the Sun and disappears, there will be a raging of lions and wolves.'*

The leading importance of the moon to all cultures from early times to the present day is undoubted, and it must be the most observed object in the sky. In fact the moon's motions, if considered in detail, are highly complicated and it has been said that it would take a good mathematician a lifetime to detect and understand them all.

Seen from earth, the moon's movements and appearances are entirely bound up with sun and earth, forming a celestial trinity which contains correspondences and rhythms which can too easily be explained away. Science has a habit of observing something then explaining it with a theory which then makes the phenomenon look inevitable. But if the observations change radically, the theory has to change too and one has the impression that whatever the observations, they will be made to fit a scheme which scales them down to serve laws of necessity, and there is nothing left to wonder at.

Whether conformable to laws or not, various details characterise a celestial body in a specific way and determine its relationship to us. Moon and sun appear approximately the same diameter in the sky, the moon keeps the same face turned to the earth, the sun's diameter is 109 times that of the earth, there are 109 sun diameters between the sun and the tip of the earth's shadow, there are 108 sun diameters between sun and earth, the earth's shadow is 108 earth diameters long, earth to moon is 111 moon diameters, the length of the moon's shadow is 111 moon diameters, the moon's lunation (e.g. full moon to full moon) is thirty days, the distance from earth to moon is 30 earth diameters, the moon's sidereal period (returning to the same star) is twenty-seven days, the rotation of the sun's equator seen from earth (synodic) is twenty-seven days . . . These times and distances are approximate figures, taken to the nearest whole numbers.

The word 'month' is connected with the Old English word 'mona' for 'moon'. Of primal interest in ancient times, and also today in certain religions, was the first appearance of the waxing crescent moon at sunset (Figure 6.1). For the Babylonians this marked not only the beginning of a month but the beginning of a day. Today the Islamic calendar is still based on the sighting of the evening crescent moon.

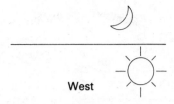

Figure 6.1

West

Sun and moon are both setting when they are at the western horizon, and as the moon's crescent is then so slender, it requires the sun to set before the moon is visible in the diminished light just above it. So one sees the first crescent moon of the month as it is disappearing below the horizon. But the moon moves eastwards with its own motion day by day faster than the sun, so the following evening the crescent moon is higher and more visible and takes longer to set. In the course of two weeks the moon will have traversed half of the celestial sphere, and its phases increase to full as it moves away from the sun to the opposite side of the horizon (Figure 6.2). These

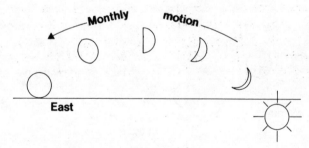

Figure 6.2

are the moon phases seen at sunset. At sunrise the other half of the moon's journey can be seen, and the phases reverse from full to waning crescent as the moon moves round its orbit to meet the sun again (Figure 6.3). A diagrammatic illustration of this seen from above the earth is shown in Figure 6.4. The light from the sun always illuminates half of the moon's sphere, which is seen at various angles from the earth, producing the phases. The sun is so distant and large that lines of sight to it are virtually parallel.

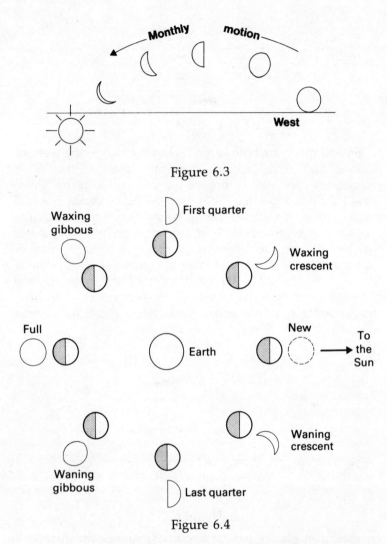

Figure 6.3

Figure 6.4

It will be noticed that, as seen from earth, the lit edge of the moon never appears to curve directly towards the sun if one imagines a straight line from the centre of the moon through the centre of the edge (Figure 6.5). The reason is that we have the impression that sun and moon lie on a sphere, or similar, and the line between them traces out a curve (section of the zodiac) on its two-dimensional surface. The line between the centre of the moon and the centre of the sun in three-

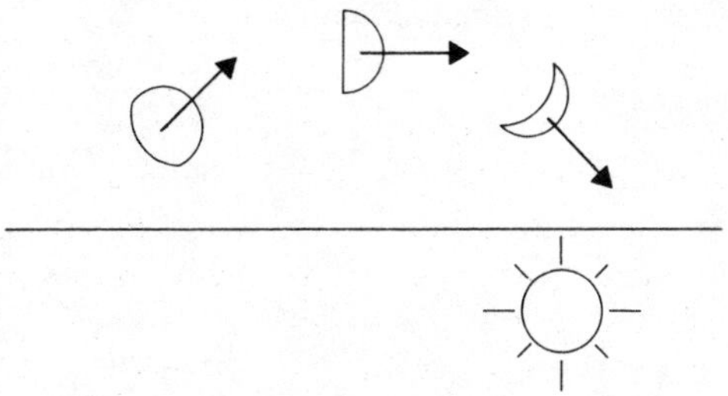

Figure 6.5

dimensional space is, in fact, straight but seems to be curved concavely to the observer. When picturing three-dimensionally it should be remembered that the moon is spherical, not a disc, and the centre of the lit edge, as seen from earth, is not necessarily the centre of the lit half of the moon. But it is an apparently curved line through the centre of the moon and its lit edge, approximately following the ecliptic, which passes through the sun – a phenomenon sometimes neglected by artists. Turner's moons are a good example of well-observed phases.

The moon's orbit is inclined just over 5 degrees to the ecliptic, so its path lies within the zodiacal constellations. This means that when the moon is full it occupies that part of the zodiac which is opposite to the sun and which the sun occupies in the opposite season – a difference of six months. Therefore, to observe the full moon in a starlit sky is as though to observe the sun in the opposite season but to see the stars it stands before. With sun and moon the same diameter, comparison is close. But the full moon's light obscures many stars.

In midwinter the full moon will rise high above the horizon and at midnight assume the position of the midsummer midday sun. More zodiacal stars are visible round the first and last quarter moons, and here one observes the equivalent of the sun's position one season removed. In spring the first quarter will show the sun's midsummer position and the last quarter will show the sun's midwinter position. The moon's nightly path above the horizon will also mirror the sun's daily path at a particular season. An indication of these sun–moon relationships is given in Figures 6.6 and 6.7. Through this it is as though the daytime stars became visible.

A consequence of the time between two full moons being

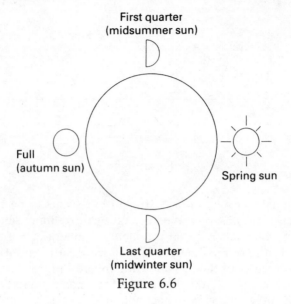

First quarter
(midsummer sun)

Full
(autumn sun)

Spring sun

Last quarter
(midwinter sun)

Figure 6.6

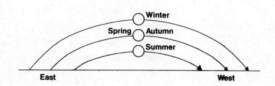

Winter
Spring   Autumn
Summer

East   West

Figure 6.7 Paths of the full moon

29½ days is that the fraction of half a day causes the exact moment of full moon to alternate between day and night from month to month. In 29½ days the sun has moved one zodiacal constellation further on, so the exact moment of full moon can, in the course of the year, occur in each constellation in succession, being a 'day' event in six of them and a 'night' event in the other six.

One characteristic of the moon's crescent phases is that their tilt to the horizon changes with the seasons. This is because the moon is always close to the ecliptic and reflects the ecliptic's changing relationships to the horizon, as demonstrated in Chapter 4. Figure 6.8 shows the waxing crescent's relationships to the western horizon at sunset and Figure 6.9 the waning

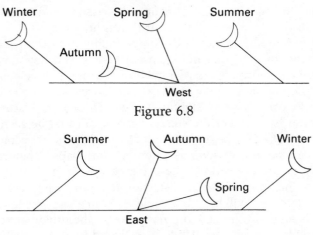

Figure 6.8

Figure 6.9

crescent's relationships to the eastern horizon at sunrise. For the waxing crescents it can be seen that in winter and summer the tilts are the same, and half-way between two extremes, while the spring crescent lies on its back (the Grail Cup moon) and the autumn crescent stands most upright. For the waning crescents, summer and winter are again the same while autumn and spring reverse the waxing crescent extremes.

If the crescents are equally distant (have the same elongation) from the sun, then greatest visibility is in spring at sunset and in the autumn at sunrise. At these times the zodiacal areas close to the sun reveal more of their secrets to earth observers and are of particular interest to the astronomer when studying the interior planets Mercury and Venus which always lie comparatively close to the sun. Spring sunset and autumn sunrise allow a deeper glance into the threshold of the 'hiding places' of stars and planets as they alternate between evening and morning appearances (see Chapter 5).

Waxing crescent visibility is still important today for the Islamic calendar which is based on the phases of the moon. Our western calendar ignores the moon and divides the solar year into twelve fixed parts. But the Islamic calendar has the first sighting of the waxing crescent as the beginning of a month. There are twelve such 'lunations' in a solar year, with approximately eleven days left over extending into the first month of the next cycle. This means that the lunar months move through the solar year and its seasons and return to a similar date in a period of about thirty-three years.

Of prime importance is the determination of the start of the holy month of Ramadan, a time of fasting and devotion. But Muslim communities have a habit of starting their celebrations on different days in any one year due to their varying locations and sky conditions for sighting the crescent moon. Places lying westwards of where the crescent moon can first be seen from the earth have a better chance of seeing it; the movement of sun and moon is westwards and more time elapses, before they arrive at the horizons of these places, for the moon to move out of the sun's light at the start of its monthly journey. Also, some latitudes will have the zodiac standing at a steeper, more visible angle to the horizon than others. Then there are delays caused by the weather. Under good conditions it is possible to sight a first crescent less than twenty-four hours after new moon (conjunction) but it is common to sight it thirty hours after. To increase the Islamic observer's chances (two independent witnesses are required), the Ottomans decreed that first sighting can be made with the naked eye *or* a telescope.

There are efforts to standardise the Islamic calendar so that Ramadan is started on the same day in the various communities. But the relationship of celestial bodies to the earth is a living thing, and every location has its own sky. So why shouldn't religious festivals begin on dates peculiar to particular places? But the modern mind wishes to generalise and abstract the situation so that the phenomena are by-passed. As with the length of the day, the average is calculated and becomes the accepted truth to accommodate the limits of circular wheels in clocks. Yet none of the celestial bodies moves in circles.

In Jewish tradition the month begins with the first crescent moon and in early times when this was sighted it was announced with a sounding of trumpets. The sighting was declared by the religious high court after hearing witnesses, and other Jewish communities were informed by fire beacons or messengers. The first day of the year, however, does not move through the seasons as does the Mohammedan, but an extra (intercalary) month is added to the years every so often to keep the beginning of the first month (Tishri) near the autumn equinox.

Here we can take the opportunity to explain the so-called

Harvest Moon as it also involves the relationship of the ecliptic to the horizon. Harvest Moons are those which rise shortly before, at and after full, on successive nights at approximately the same time so that when the light of the setting sun diminishes in the west, the light of the moon takes over from the east. This happens near the autumn equinox when the ecliptic at sunset lies low along the eastern horizon (Figure 6.10). There-

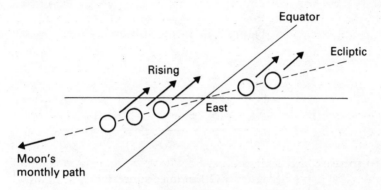

Figure 6.10

fore the moon's monthly movement does not carry it very far below the horizon at sunset from night to night, and its rising times are close to each other around sunset. The time difference on consecutive nights can, at the latitude of London, be around twenty minutes at Harvest Moon, while in spring the moon can rise 1 hour 20 minutes later on consecutive nights. The full moon following one month after Harvest Moon is called the Hunter's Moon when the quick consecutive rising effect is similar but decreased.

It will be noticed that on different nights the moon rises at different points along the horizon. In one month the moon will rise and set roughly at the places which the sun occupies on the horizon in the course of a year. In addition, the moon's path in the sky in a month will imitate the sun's path, high or low, in a year. But one particular phase of the moon will take a year to rise, pass over, and set in imitation of all the sun's annual motions. This follows from what was discussed in connection with Figures 6.6 and 6.7. It results in every season having its own individual pattern of lunar phenomena. Figures 6.11 to 6.18 show moon phase risings, diurnal paths, and

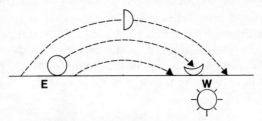

Figure 6.11 Spring sunset

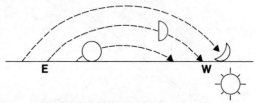

Figure 6.12 Summer sunset

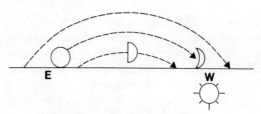

Figure 6.13 Autumn sunset

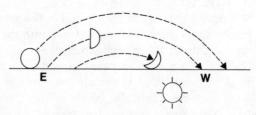

Figure 6.14 Winter sunset

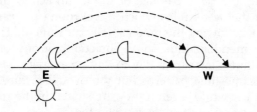

Figure 6.15 Spring sunrise

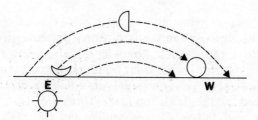

Figure 6.16 Summer sunrise

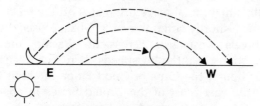

Figure 6.17 Autumn sunrise

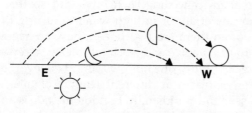

Figure 6.18 Winter sunrise

settings for the four seasons.

The rising and daily passage of the moon across the sky is reflected in the ocean tides which travel round the earth from east to west, following the moon's daily movement. The moon crosses the meridian (due south direction) of any one place about 50 minutes later from day to day on average. Most tides follow the same time pattern. But the overall picture of tides throughout the world, even if they are similar in the rhythm of their occurrence, is extremely complicated.

Leaving aside a theoretical approach to the moon, the earth, and the earth's envelope of water – a major phenomenon is that the earth's main tidal wave originates off the west coast of South America shortly after the moon has passed overhead there. It travels westwards across the Pacific, moving in different directions at different speeds. It reaches Australia in about fifteen hours, the Cape of Good Hope in about twenty-nine hours, the east coast of the United States in about forty hours, the west coast of Ireland in about forty-four hours, and round the north coast of Scotland down to London in about sixty hours.

This tide started when the moon was overhead off South America, and when the moon is directly 'underneath' on the opposite side of the earth 12 hours 25 minutes later, a second tide starts across the Pacific from South America. At any one time there are about five or six tides travelling round the earth and separated by 12½-hour intervals. Tides are also set up in the Atlantic and Indian Oceans which join the main Pacific tide. Tides also approach a coast from opposite directions (this happens in the North Sea) and in some parts of the world this can result in them cancelling each other out so that there is practically no tide at all, or in high water occurring four times a day. More often there are two tides a day, an average of 12 hours 25 minutes apart, and each of these is repeated the following day about fifty-one minutes later. However, tides take place at different places at different lengths of time after the moon has passed the meridian. In London the tide is an average of just under two hours behind the moon, while at New York the tide is an average of 8¼ hours behind. The astronomical clock above a gateway to Hampton Court Palace on the Thames, built in the reign of Henry VIII, still indicates the time the moon crosses the meridian every day in the determining of tides.

Spring tides have nothing to do with the spring season, but the term comes from the Old English word *'springan'*, meaning to rise. They are the highest tides of the month and are associated with full and new moons. The highest spring tides take place in connection with the full or new moon coinciding with the moon's perigee (the time when the moon is nearest to the earth) and with the earth's perihelion** (when the sun is nearest to the earth – i.e. in January). January is also a time when strong winds can raise the tide above normal. Neap tides are the low tides of the month occurring between the two spring tides and associated with first and last quarter moons. Highest spring tides have been recorded as occurring in a cycle of 18.6 years, which is in accordance with the movement through the zodiac of the moon's perigee and node, these latter movements to be discussed in the chapter on eclipses (Chapter 7).

Tidal patterns throughout the world are so complicated that they allow one place, the island of Tahiti, to have high tides which occur at almost the same times each day (noon and midnight) and neap tides at 6 a.m. and 6 p.m. Here the influence of the sun is predominant and for some reason the rhythm of the moon does not manifest.

The heavens move in their cycles and the earth responds, taking up the rhythms in its own way in a particular place. It is interesting to note that the place of origin of the tides, the Pacific, where the ocean most strongly responds to the moon's presence, is also the place of origin that has been claimed for the moon itself. It has been suggested that at an earlier stage of evolution the moon was part of the earth and that the earth's molten liquids came under the influence of the sun in huge tides which separated the moon from the Pacific area and threw it into orbit (fission theory). Scientists are not agreed on the origin of the moon, two other theories being that it was a separate body in space which was 'captured' into an earth orbit; or that earth and moon formed together as a 'double planet'.

The Apollo series of American spacecraft brought samples of moon rock to earth for analysis and in *The Moon Book* Bevan French sums up –

Despite the flood of chemical and historical information obtained by the Apollo Program, we still do not have a single, universally accepted theory for the origin of the moon. Because scientific theories always die hard, the three pre-Apollo theories (double planet, fission, and capture) have all survived the Apollo results, though often with considerable modifications. A completely successful theory of lunar origin . . . must account for significant differences in the chemistries of the earth and the moon. This chemical disparity is the major stumbling block of the 'double planet' theory, which argues that the earth and moon were formed together in the collapsing dust cloud that became the solar system . . . A newer variation of the fission theory suggests that the moon was built up gradually from a heated atmosphere that was thrown off a hot, rapidly spinning primordial earth . . . As the atmosphere cooled, the less volatile elements condensed into small rocky particles which were spun into orbit around the earth and then assembled to become the moon.

One problem with the 'capture' theory is that the captured body has to be slowed down in order to go into orbit around a planet like the earth. The earlier tidal fission theory, originally propounded by G. H. Darwin, son of Charles Darwin, in 1880, would now require the moon to have been a pocket of somewhat differently composed material to the rest of the earth when it was flung off. In addition, if it is to be tenable that the moon separated from the region of the Pacific, then, in the light of recent research, it must have had a constitution such that, on analysing it today, it appears to be of a greater age than the floor of the Pacific, as has been revealed by deep-sea drilling. Because one is dealing with earlier, unknown states of matter, to draw final conclusions from present research is not easy. That the moon came out of the Pacific was proposed by W. H. Pickering (1858–1938), the American astronomer who discovered Saturn's ninth satellite, Phoebe. Whatever the conclusions of the scientific world, the Pacific waters continue to respond to the moon by setting up the earth's primary tidal rhythm.

The influence of the moon on the earth has long been a subject of debate. Plant growth and weather are two prominent areas in the discussion and it is difficult for science to provide experiments which are sufficiently sensitive and prolonged. The researcher Agnes Fyfe in Switzerland tackled this problem using a filter-paper method of 'capillary-dynamolysis' on plant sap over a period of about twelve years, involving over 70,000 tests. The results were positive, in favour of a correlation between sap activity and the moon, as described in her book *Moon and Plant*. The rise and fall of tides and plant sap are not, essentially, separate phenomena. Creatures which live in the sea, such as oysters, sea urchins, worms, and grunions, also respond to the moon rhythms, as described, for example, by Rachel Carson in the chapter on tides in her book *The Sea Around Us*.

An article in *Nature* by Peter Kahn of Princeton University and Stephen Pompea of Colorado State University in 1978 links rhythms of the moon with fossilised nautilus shells, suggesting that there was a time when the moon was about half the distance from the earth than it is now and that it orbited the earth in about one-third of the time it takes now. The present-day astronomer understands that the moon is drawing away from the earth at a rate of about 3 metres a century. The physical dynamics of the earth–moon system also require that the moon is, at the same time, speeding up its motion. According to calculation, this would end up with the moon orbiting the earth in the same time it takes the earth to rotate, so that the moon would hang motionless in the earth-observer's sky. V. A. Firsoff, in his *The Old Moon and the New*, takes the calculation further to the stage when the moon might return towards the earth, adding 'I will not pursue this cosmic Ragnarok*** to its catastrophic conclusion'.

The above considerations on lunar science are included in this exposition on observational astronomy as the moon, being the closest of the 'planets', has since earliest times lent itself to the impression of being a physical body as well as a celestial one. Unlike the more distant planets, its 'face' is visible and its physiognomy available to be read as one wishes. A strange, pre-Christian quotation places Britain in a special role concerning the physical and celestial faces of the moon. In the first century BC, the historian Diodorous of Sicily wrote how it was said that, in Britain,

> [T]he moon, as viewed from this island, appears to be but
> a little distance from the earth and to have upon it
> prominences, like those of the earth, which are visible
> to the eye. The account is also given that the god visits

the island every nineteen years, the period in which the
return of the stars to the same place in the heavens is
accomplished; and for this reason the nineteen-year
period is called by the Greeks the year of Meton.

(See Chapter 7 for a description of the 'Metonic Cycle'.)

Plutarch (AD 50–120), in his celebrated discussion *On the Face
in the Round of the Moon* which influenced both Copernicus and
Newton, has one contributor to the conversation say: 'It is not
probable that the moon has but one superfices all plain and
even, as the sea; rather, that of its nature it principally resembles
the earth.' On the other hand, he adds: 'And we are very far
from thinking that the moon, which we hold to be a heavenly
earth, is a body without soul and spirit, exempt and deprived
of all that is to be offered to the Gods.' This conversation on the
moon includes references to earlier astronomers like Aristarchus
and consideration of the earth's orbital movement, and to the
germinal idea of the moon moving according to the laws of
gravitation. Yet at the same time the discussion concludes with
a description of one of the moon's functions as that of a staging
area for departed souls before they move out into the cosmos.
The moon has stood in the sky as an intermediary, a stepping-
stone on the border leading to the materialisation of the
universe on the one hand and to the hierarchies on the other.

In the ancient east the moon had its own zodiacal divisions.
Among peoples like the Chinese, Indians, Persians, Arabs and
Copts, a system of dividing the ecliptic into twenty-eight (or
twenty-seven) parts was used, within one of which the moon
would remain for about a day. For the Chinese they were
'mansions', for the Indians they were 'wives of the moon' and
the Arabs called them 'alighting stations'. Some of the divisions
were marked by major zodiacal stars including the Pleiades,
Aldebaran, Betelgeuse, Castor and Pollux, Regulus, Spica, and
Antares. A twenty-fourth century BC Chinese chart of the
mansions shows the divisions as unequal but organised into
four groups which start and finish in alignment with the four
seasonal points of the sun. In addition, the Chinese mansions
derive their areas from measurements made along the celestial
equator, not the ecliptic, thus indicating a system related essentially to the celestial pole. Also, the divisions as used by
different cultures do not always coincide with each other,

though nine 'boundary' stars in the Chinese system are the same as those in the Indian.

The precise movement of the moon against the stars is, as said before, extremely complicated. However, two fundamental aspects for ordinary observation are the positioning of the phases against the zodiac, and the surface area of the moon which is revealed at these phases.

Firstly, when the moon has moved from conjunction with the sun to first quarter, it stands more than 90 degrees round the zodiac from the conjunction point. The reason is that the sun moves constantly through the zodiac in the same direction. It therefore takes longer for the moon to move 90 degrees from the sun than it does to move 90 degrees from a star. Subsequent phases, linked to the sun, increase the disparity, until the next new moon occurs at the sun's position a month later and about one constellation further round the zodiac (Figure 6.19). This expresses the difference between a synodic month and a sidereal month.

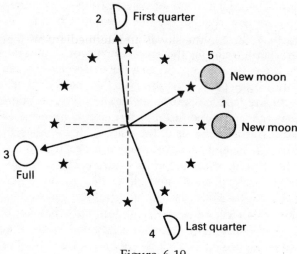

Figure 6.19

Secondly, more than half of the moon's surface is visible from one place on earth in the course of a month. The moon appears to dance gently against the stars, nodding vertically and swaying horizontally through the zodiac. The nodding is 'libration 'in latitude' and occurs because the moon keeps its poles in a constant direction in relation to the stars while orbiting the earth, so that now more, now less is seen of its north and south areas (Figure 6.20).

Earth

Figure 6.20

The horizontal swaying, or libration in longitude, takes place because the moon is moving on an ellipse, not a circle. The earth is at one focus of the ellipse and when the moon approaches this (the moment of perigee) it moves faster. But the turning of the moon on its axis does not change, so the two movements fall out of step. The axial movement is synchronised with the earth, and the moon would keep one unchanging face to the earth if it moved in a circle. But with the speeding up of the orbital movement on an ellipse, the face which would otherwise be turned to the earth is swept on past the earth before it can turn sufficiently, so to speak. The observer then sees further round one side of the face as it turns away slightly. The interesting outcome is that the same face of the moon is found always to be turned towards the focus of the ellipse which the earth does not occupy. This is the 'focus' of the moon's rotation on its axis. The earth is the focus of its orbit. Figure 6.21 exaggerates the ellipse for the sake of clarity and places new moon at perigee – though perigee can coincide with any phase.

Maximum libration occurs twice a month at whatever phase, each occasion revealing an area on the opposite edge of the face to the other (Figure 6.22). At perigee and apogee the libration effect does not occur as the moon is turned towards both foci at once. Over a number of years (there are many other move- ments to be considered) up to 59 per cent of the total surface of the moon can be mapped from earth.

In legend, a creature frequently associated with the moon is the hare, which figures in stories from diverse cultures.

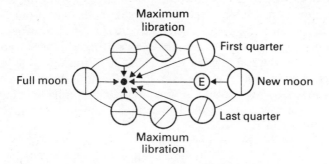

Figure 6.21

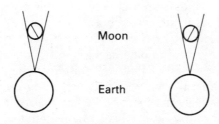

Figure 6.22

According to a Hindu legend, Buddha was a hare in an early stage of his existence. He travelled in the company of an ape and a fox and one day they were met by a beggar who asked for food. The hare alone was unsuccessful in finding anything for the beggar but instead had a fire built and cast himself into it to roast for the man's supper. The beggar was the god Indra in disguise and had the hare placed on the moon for his sacrifice. Likewise the moon gives itself up monthly to the light or fire of the sun at conjunction. In Sanskrit the moon is called Sasanka – 'having the marks of a hare'. The hare is also associated with the moon in China and South Africa, and in Mexico the moon's face marks are said to be those of a rabbit. Taoist fable depicted a hare pounding the drugs which compose the elixir of life, and the Chinese also represent the moon by a rabbit pounding rice in a mortar. It has been noted in literature on moon mythology that the rabbit has a gestation period of thirty days – comparable to the synodic month.

One's little finger held up at arm's length covers the face of the moon in the sky. Yet this celestial body has prompted a vast amount of literature, imaginative and scientific, throughout history. It is the closest neighbour to earth and moves round it assured of the response of nature and man from tides to calendars. It stands in the sky as a stepping-stone to the heavens beyond, whether physical or imaginative, faithfully keeping the same face turned towards the earth as if linked by an invisible umbilical cord.

What if the moon were not in our sky? This question is investigated in an essay by the science writer Isaac Asimov. His opening words are:

> There was a full moon in the sky this morning. I was awake when the dawn had lightened the sky to a slate blue (as is my wont, for I am an early riser) and, looking out the westward window, I saw it. It was a fat yellow disk in an even, slate-blue background, hanging motionless over a city that was still dreaming in the dawn. Ordinarily I am not easily moved by visual stimuli, as I am relatively insensitive to what goes on outside the interior of my skull. This scene penetrated, though.

Let it be hoped that the visual stimulus of the moon continues to penetrate the skull of modern man and draw him out to experience directly the phenomena of the heavens.

* It appears that the moon reaching the sun means new moon or conjunction; that the moon and sun being seen together on the 14th means that the near-full moon rose before the sun set; and the moon not waiting for the sun means that when near-full it sets before the sun rises.

** This perihelion point (an apse of the earth's orbit) makes one eastward revolution round the zodiac in about 110,000 years.

*** Twilight of the gods, or day when the world will be annihilated before being later reconstructed on an imperishable basis.

Chapter 7
Eclipses

The phenomenon of eclipse is that of light being darkened over and above ordinary day-to-day changes of light and darkness. It is as if an intervention took place in the usual affairs of nature. Yet the interventions, if examined closely, have a remarkable rhythm of their own which Man has studied and tried to analyse since very early times.

Twice a year there are eclipse periods. During these times the light of the sun and moon is diminished or well-nigh extinguished. At least four eclipses occur in any one year, though two of these (darkenings of the moon) may not be visible to the naked eye. At least two visible darkenings of the sun must take place within twelve months. A maximum of seven normally visible eclipses can occur in a year – five of the sun and two of the moon, or four of the sun and three of the moon.

The eclipse periods are opposite each other in the year, about six months apart. These periods are called 'eclipse seasons'. They rotate once through the four seasons of the year in 18½ years.

Sun eclipses are awesome spectacles, even for the modern, unsuperstitious observer. One is conscious of a gigantic event of nature taking place, over which the human being has no control. With a total sun eclipse the day-time sky darkens into an eerie light with a reddish glow round the circle of the horizon. Bright stars shine out, day-time birds and animals fall silent and flowers and leaves close up which normally only close at night. The atmosphere grows cold and dew may fall. The sun becomes a ring of diffuse light in the sky, rose-coloured around the edge of a dark disc. This can last for a maximum period of about 7½ minutes (near the earth's equator) and just over six minutes at a latitude of 50 degrees. The return of the sun to visibility is with a sudden flash of light across the land-scape. The time of the whole process from the commencement of darkening to full restoration of light can approach four hours.

A dramatic diminution of light occurs just before total sun eclipse and unless one is watching for it the earlier stages can pass unseen. The sun, even as a slender crescent in a clear sky, is very bright, and it has been estimated that a 98 per cent eclipse in clear conditions could go unnoticed. Yet thin cloud cover over even a small partial eclipse makes it readily visible. Turning our attention to the earth at our feet, we can notice that the light filtering down through trees throws patches of light on the ground which are crescent-shaped, arising from projections of the sun's crescent shape through narrow open-ings. This is on the principle of the camera obscura and was observed and commented upon by Aristotle. It should be added that a projection of the sun's image onto a screen is the safest way of observing a solar eclipse to avoid eye damage. Certainly the sun must never be looked at with the unprotected eye, and professional advice should be sought on the use of darkened glass, etc.

As to eclipses of the moon, the complete process of a total lunar eclipse from the first faint darkening (not visible to the naked eye) can take up to six hours. As the eclipse progresses the moon frequently reddens or takes on a copper colour. On rare occasions it can disappear completely from the sky. From beginning to end the readily-visible darkening can take almost 4 hours. The length of the middle stage of greatest darkening (total eclipse) can last almost 1¾ hours.

The word 'eclipse' comes from a Greek word rendered vari-ously as 'failure', 'abandonment' or 'disappearance'. From it is derived the term 'ecliptic' – that is, the circle on the celestial sphere on or near which eclipses of sun and moon occur. Early peoples spoke of a celestial dragon which caused the darken-ings, and in old illustrations it is shown with its head at one eclipse area in the zodiac and its tail at the other, on the opposite side of the zodiac. According to Babylonian myth, the god Marduk created this Great Dragon, making it carry six of the zodiacal constellations on its back and six under its belly.

In India, people immersed themselves in water up to their necks during eclipses and in Japan wells were covered to prevent poisoning. In arctic America the Eskimo, Aleut and

Tlingit peoples said that during an eclipse the sun or moon temporarily left the sky to see that things were in order on earth. Plutarch reports how, at a total moon eclipse on the eve of the battle of Pydna, '[t]he Romans, according to their custom, made a great noise by striking upon vessels of brass and held up lighted faggots and torches in the air in order to recall her light.' Tacitus mentions an eclipse of the moon as happening soon after the death of Augustus when soldiers 'made a loud noise, by ringing upon brazen metal, and by blowing trumpets and cornets; as she appeared brighter or darker they exulted or lamented'. In medieval European chronicles it was common to describe the moon as 'turned into blood' when it took on a reddish colour when eclipsed.

Eclipses have been long looked upon as events of ill omen or change. Also, there is a tradition that they are connected with earthquakes. The Book of Revelation has been cited as an example; it says, 'And I beheld when he had opened the sixth seal, and, lo, there was a great earthquake; and the sun became black as sackcloth of hair, and the moon became as blood.' Thucydides states that during the Peloponnesian war 'things formerly repeated on hearsay, but very rarely confirmed by facts, became not incredible, both about earthquakes and eclipses of the sun which came to pass more frequently than had been remembered in former times'.

Early Chinese records make it clear that astronomical observations, including those of eclipses, were for the purpose of political astrology on behalf of the emperor and his family. The record of a near-total eclipse, on 18 January AD 120, states: 'There was an eclipse of the sun. It was almost complete and on the earth it became like evening . . . The woman ruler showed aversion to it. Two years later, Teng, the Empress Dowager, died.' The Chinese character for eclipse totality (*chi*) was originally the pictogram of a man turning his head away from a plate of food, signifying that he was satiated.

As far as existing records are concerned, Babylonian astronomers made accurate observations of celestial phenomena, including eclipses, since 750 BC. This is shown in Babylonian astronomical tablets which were accidentally dug up last century by people looking for ancient, baked-clay bricks for their buildings. Most of these tablets are now in the British Museum, but all of them represent only about 5 per cent of the original archive. Older texts, dealing with astrology, include the extensive collection of omens called *Enuma Anu Enlil* which appear to date back to before 2000 BC. A characteristic of these old texts is that, in eclipses of the moon, the four quarters of the face of the moon represented the four countries of the land. A partial eclipse over one of the quarters indicated an omen for the appropriate territory, with a total eclipse related to all the countries. Particular territories were also indicated by the timing of the eclipse in terms of month, day, and hour. In this way astrology led to the necessity for an accurate science of observational astronomy.

The basic phenomenon of a solar eclipse is that the sun appears to pass through 'phases', some of which resemble those of the moon when crescent-shaped. As a total eclipse of the sun can take almost four hours, the sun will move in diurnal motion westwards in this time, so Figure 7.1 must be read from left to right. However, the effect is caused by the moon passing across the face of the sun from right to left as part of its own monthly motion (Figure 7.2).

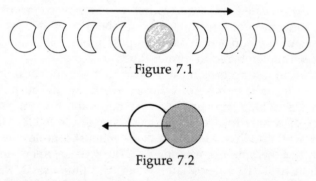

Figure 7.1

Figure 7.2

A total lunar eclipse has an opposite appearance. Here the diurnal movement is again from left to right (Figure 7.3) but the darkness is caused by the moon entering the earth's shadow from right to left (Figure 7.4).

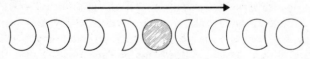

Figure 7.3

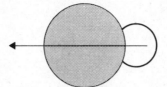

Figure 7.4

The technical reason for these phenomena is the afore-mentioned remarkable geometrical relationship between sun, earth and moon (Figure 7.5). Seen from the surface of the earth,

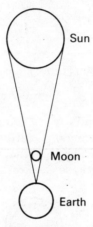

Figure 7.5

the diameter of the moon is approximately the same as the diameter of the sun. When the moon is new and also stands exactly between earth and sun, there is an eclipse of the sun. When the moon is full and sun, earth and moon are also in line in space, there is an eclipse of the moon (Figure 7.6). The drawings are diagrammatic to show the geometry of the situation. In an eclipse of the sun, the moon overtakes and passes directly in front of the sun; in an eclipse of the moon, the moon overtakes and enters the earth's shadow.

The distances between sun and earth and moon and earth vary rhythmically, giving rise to three basic types of solar eclipse. For instance, when the moon is closest to the earth (perigee) it will appear larger than the sun and if an eclipse occurs during this time, the disc of the moon will appear larger

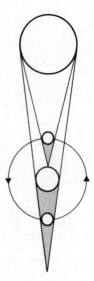

Figure 7.6

than that of the sun and cover it completely, resulting in a total solar eclipse. When the moon is furthest from the earth (apogee) its disc will appear smaller than that of the sun and an annular (Latin: 'little ring') eclipse can result (Figure 7.7). Here the

Figure 7.7

shadow of the moon is not long enough to reach the earth (Figure 7.8), sometimes being short by more than a moon diameter. Thirdly, when the observer is not directly within the moon's shadow but to one side of it, a partial eclipse occurs for him (Figure 7.9).

As the moon passes across the face of the sun from west to east, this causes its shadow to move across the surface of the earth from west to east, a journey which can take in the region of six hours at a ground speed of over 1,600 kilometres per hour. Geographically the shadow traverses less than the semi-circumference of the earth as the earth is rotating in the same

Figure 7.8

Figure 7.9

direction as the moon's shadow. In space the shadow travels at about twice its ground speed on earth. Its maximum diameter on the surface of the earth when perpendicular to it is about 250 kilometres. Figure 7.10 shows the general direction of the shadow's movement.

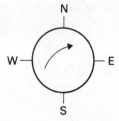

Figure 7.10

A solar eclipse makes its first appearance at those places geographically westwards experiencing sunrise, and will last be seen from places eastwards experiencing sunset (Figure 7.11). Places at the western limits of eclipse visibility will view the end of an eclipse as the sun rises, and places at the eastern

Figure 7.11

limits will view the beginning of an eclipse as the sun sets. At these two extremes the least part of an eclipse is seen. Between these extremes, eclipses will be experienced at times of fore-noon, noon, or afternoon, depending on the location. There are also northern and southern limits, about 3,500 kilometres distant from the shadow path, within which the observer sees a partial eclipse and at the extremities of which an eclipse ends the moment it begins.

The shadow of the moon is called the umbra (Latin: 'shadow') and surrounding it is an area part-light and part-shadow called the penumbra (Latin: 'almost shadow') – Figure 7.12. An

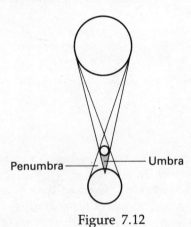

Figure 7.12

observer within the moon's umbra will see a total solar eclipse and from within the moon's penumbra he will see a partial solar eclipse.

The earth too has its umbra and if the moon stands within it the observer on that side of the earth which faces the moon will see a total lunar eclipse (Figure 7.13). If the moon passes only

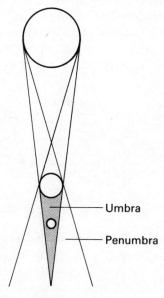

Figure 7.13

partly through the cone of the earth's shadow, a partial lunar eclipse will be seen from earth (Figure 7.14). If the moon misses the earth's shadow but enters the earth's penumbra, then a penumbral lunar eclipse will occur in which the face of the moon will only darken slightly, sometimes not visibly to the naked eye.

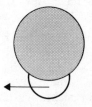

Figure 7.14

So, solar eclipses can be total, annular or partial; and lunar eclipses can be total, partial or penumbral.

Normally, at new moon the moon's shadow misses the earth (passes to north or south of it in space) and there is no solar eclipse; and at full moon the moon normally misses the earth's shadow and there is no lunar eclipse. But there are two periods of the year when the situation is different and sun, moon and earth can come into line in three-dimensional space. This occurs at the moon's nodes (Latin: 'knots') where the yearly path of the sun and the monthly path of the moon cross, as seen from earth (Figure 7.15). The diagram shows new moon when there

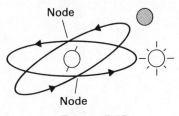

Figure 7.15

is no eclipse as sun and moon stand apart in the sky. But if they stand together near one of the nodes, within certain limits, there is a solar eclipse situation and if the sun is at one node and the moon at the other, there is a lunar eclipse situation as in both cases sun, earth and moon can lie in a line. The place where the moon crosses the ecliptic from below upwards or from south to north, is called the ascending node, and the place where it crosses the ecliptic from north to south is called the descending node. The ascending node has the symbol ☊ and was designated the Dragon's Head in early times, and the descending node with the symbol ☋ was designated the Dragon's Tail.

In one year a maximum of seven eclipses are possible, excluding lunar penumbral eclipses. This is what was meant by seven 'normally visible' eclipses in a year mentioned on page 87. An example of how these could be placed is shown in Figure 7.16. The eclipses, taking place within twelve new moons, are numbered in order of occurrence with 1, 3, 4, 6 and 7 as solar eclipses and 2 and 5 as lunar eclipses. A solar eclipse must happen when new moon occurs within 15 degrees 23 minutes of a node, and a lunar eclipse (not penumbral) must happen if

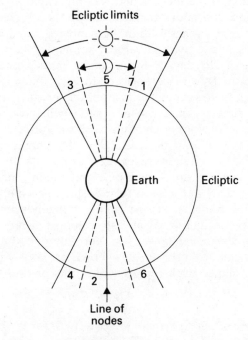

Figure 7.16

full moon occurs within 9 degrees 39 minutes of a node.

Further, these nodal points move westwards round the ecliptic, making a complete rotation against the stars in 18.6 years. Therefore in nine years, eclipses of sun and moon will be spread round the whole zodiac. The movement of the nodes by about two-thirds of a constellation per year is caused by the turning of the plane of the moon's orbit in space (Figure 7.17). This also means that the sun, moving eastwards on the ecliptic, will meet a particular node in less than a year – about 18.6 days

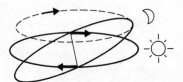

Figure 7.17

less in fact. This amounts to 346.62 days, known as a synodic revolution of the moon's nodes or an eclipse year.

As said, the nodes precess westwards completely round the ecliptic in 18.6 years. Within these moving nodal regions dozens of eclipses take place at various points, but in a special period of 18 years 11 1/3 days (referred to as a Saros period) two eclipses of the sun take place adjacent to each other in relation to a node, and likewise two eclipses of the moon. The reason is that in this time (6,585 days) one node will have made a whole number (nineteen) of meetings with the sun (eclipse years) and the moon's synodic periods (full or new moons) will also have completed a whole number (223) of circuits in relation to the sun. Therefore the sun, moon and moon's node will stand almost exactly in the same relationship to each other in the sky.

There is a further lunar motion, strangely synchronised with the above, which adds an astonishing finishing touch to the these adjacent Saros eclipses. Not only does the plane of the moon's elliptical orbit turn as shown previously in Figure 7.17, but the ellipse itself rotates within this plane (Figure 7.18). It

Figure 7.18

does so anti-clockwise, or eastwards seen from the earth, taking about nine years to complete a rotation. The effect is that the point of perigee, for instance, moves eastwards through the zodiac constellations, making one circuit in nine years, two in eighteen years. Therefore, when two Saros cycle eclipses take place adjacent to each other in the sky, the apparent sizes of sun and moon are approximately the same and there is a near repetition of the eclipse phenomenon 18 years 11 1/3 days earlier (the exact whole number of days depending on the number of leap years between).

In relation to the node, the repeated eclipse of sun or moon will take place about one sun (or moon) diameter further west-wards. To take the case of the solar Saros cycle, these eclipses will step through the entire nodal region in a period of between 1,244 and 1,515 years. This stepping is diagrammatically shown in Figure 7.19. The first eclipse (1) of this series at the ascending

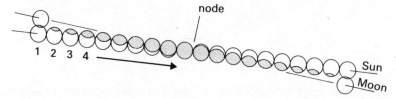

Figure 7.19

node will be visible as a small partial solar eclipse seen only from the earth's north pole region because the obscuring moon is north of the sun in latitude (Figure 7.20). The next eclipse in

Figure 7.20

the series will bring a shadow path further south on the earth but about one-third further round the globe. This is because the eclipse occurs 18 years 11$\frac{1}{3}$ days after the last one, so the earth will have rotated one-third further on. So consecutive eclipses move progessively southwards and westwards when at the ascending node, spiralling clockwise round the earth from north pole to south pole in, as said, a period of between 1,244 and 1,515 years (Figure 7.21). During this time between seventy and eighty-five solar eclipses occur. At the descending node the first solar eclipse of a series appears as partial at the south pole.

As can be seen from Figure 7.19, the eclipses in the middle of a series become total or annular. This stage lasts for about forty or more eclipses, or over two-thirds of the series. At any

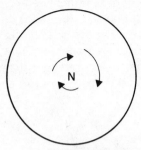

Figure 7.21

one time there are about forty solar Saros series in the process of spiralling southwards or northwards. On average a new eclipse series begins every thirty-one years, and series come to an end at the same rate. In a Saros period of 18 years 11¹/₃ days, an average of 86.4 eclipses, comprising solar and lunar events (including lunar penumbral) take place, involving both nodes. If one excludes the lunar penumbral events, the average drops to 70.6. Because of the special geometrical relationships involved, for the area of the whole earth there are roughly the same number of total solar eclipses occurring as there are annular eclipses – the frequency being once every 1.4 years on average for each type.

Looking at eclipse phenomena against the background of the stars, it is found that successive eclipses in a Saros cycle (adjacent to each other in relation to the node) take place roughly one-third of a constellation apart, progressing eastwards. This is an expression of the eleven days in excess of a whole number of years of the Saros period. (From year to year, new moons or full moons which are separated from each other by twelve synodic months also step through the zodiac approximately one-third of a constellation apart – though westwards – explaining the difference between a lunar year and a solar year: i.e., about eleven days which is also the fraction involved in the length of a Saros period.)

The term 'Saros' commonly used for this period of 18 years 11¹/₃ days (6,585.32 days) is a Greek word with the Babylonian root 'Sar'. However, the Babylonians never used this term in connection with this period, instead sometimes calling it 'the eighteen'. In fact, the word 'Sar' related to a period of 3,600 years. But 'Saros' is now part of the astronomer's vocabulary concerning eclipses.

The Babylonians are also often credited with the discovery of the eighteen-year Saros to predict eclipses. But it is not feasible that this referred to solar eclipses as Babylonia is a small area on the globe and successive solar eclipses in a Saros are only visible from different longitudes and latitudes as their paths spiral round the earth. From any one point on earth (not an extended area) an average of one total or annular solar eclipse is visible very 140 years. Total solar eclipses are only seen on average once every 375 years from one point.

The mean frequency of total solar eclipses is highest in the northern hemisphere due to the earth being furthest from the sun (aphelion) in July when the sun spends more time above the horizon in the north and when the moon's shadow is longest. As to the moon's closeness to the earth, it is worth, in passing, to note the extraordinary fact that 'extreme' perigees and apogees take place only during the winter in the northern hemisphere when the earth is nearest to the sun; and that these extremes (say of perigee) have a habit of occurring at intervals of 18 years 11 days – the Saros period! Yet these events are not involved with eclipses. For the period AD 1750 to AD 2125 the closest approach of the moon to the earth was on 4 January 1912. That night was also full moon (not eclipsed) and with the earth's perihelion occurring the day before, the moon would have been slightly brighter than usual. The furthest distance between earth and moon for this period occurred on 2 March 1984 – also the furthest to be reached until the first quarter of the twenty-second century AD.

However, concerning the Saros period, the Babylonians can be credited with discovering or knowing about 18-year 11-day intervals between lunar eclipses, the reason being that a lunar eclipse is visible from an area of about half the surface of the earth – that half turned towards the moon when it enters the earth's shadow. Therefore there are a greater number of visible lunar eclipses from any one place than solar – about twice as many, even excluding lunar penumbral.

Because a Saros period has one-third of a day as one of its factors, every third eclipse in a series will occur at roughly the same time of day as well as its being roughly in the same place in the sky, and in the case of lunar eclipses this allows them to be experienced from one location regularly every $18 \times 3 = 54$ years. The Greek writer Geminos (c. 70 BC) ascribes this period to the Chaldeans and it is also mentioned in a cuneiform text from Uruk. Greek authors called this period *'exeligmos'* which means 'full rolling-off'. There are, in addition, many references to an eighteen-year period for lunar eclipses in late Babylonian times – from the reign of the Persian King Cyrus (539 BC) onwards. A lunar Saros cycle (of umbral eclipses only) lasts for between 685 and 1,046 years, containing from thirty-nine to fifty-nine eclipses.

Knowledge of the lunar Saros period in ancient times is remarkable, considering the limited geographical area which a particular culture covered and the amount of systematic observation and calculation required to detect regularities. With the development of modern, global consciousness, one can now speak of phenomena like eclipses as they occur for the whole earth. For example, the surface of the earth experiences the same number of solar eclipses as it does lunar eclipses to a high degree of accuracy – if one includes those penumbral lunar eclipses which are faint enough to escape the naked eye. This works out at an average of 2.38 solar eclipses and 2.41 lunar eclipses in one year. On this basis there is at least one solar and one lunar eclipse in each eclipse season or period. However, if the faint penumbral lunar eclipses are not included, then it is possible for there to be no lunar eclipse during a calendar year. On average, for the whole surface of the earth, there are twenty solar eclipses to thirteen visible (non-penumbral) lunar ones. Lunar eclipses which are not penumbral have an average frequency of only 1.54 per year.

Again considering the whole earth, a rare type of solar eclipse can be detected which is a combination of total and annular. At the beginning and end of the shadow track the eclipse is seen as annular, but in the middle where the surface of the earth is closer to the moon it becomes total. These are called annular-total or central eclipses and they occur about twice in a Saros cycle. There was one in April 1912, just total in Portugal, and its successor was just total in California in April 1930. It appears that in early times a distinction between total and annular eclipses was not made, both types being regarded as total. Distinction between them begins to occur only after about AD 1000.

Today the calculation of where and when eclipses took place in early times can sometimes help to ascertain the dates of historical events in connection with which eclipses were mentioned in contemporary texts. A well-known example which, however, has had a history of some controversy among modern scholars, is provided by Herodotus who writes about the prediction of a solar eclipse by Thales in connection with a battle between the Medes and the Lydians. He says:

> As the balance had not inclined in favour of either nation,
> another engagement took place in the sixth year of the
> war, in the course of which, just as the battle was
> growing warm, day was suddenly turned into night.
> This event had been foretold to the Ionians by Thales of
> Miletus, who predicted for it the very year in which it
> actually took place. When the Lydians and Medes
> observed the change they ceased fighting and were alike
> anxious to conclude peace.

Modern calculations show that a total solar eclipse occurred on 28 May 585 BC, at the place of the battle in northern Turkey.

An example of the dating of a piece of writing concerns *Della Composizione del Mondo* by Ristoro d'Arezzo, which states:

> And while we were in the city of Arezzo, where we were
> born, and in which we are writing this book, in our
> monastery . . . one Friday at the sixth hour of the day,
> when the Sun was 20 degrees in Gemini, and the
> weather was calm and clear, the sky began to turn
> yellow, and we saw the whole body of the Sun covered
> step by step, and obscured, and it became night; and we
> saw Mercury close to the Sun, and all the stars which
> were above the horizon; and all the animals and birds
> were terrified; and the wild beasts could easily be caught
> . . . and we saw the Sun entirely covered for the space
> of time in which a man could walk fully 250 paces; and
> the air and the ground began to become cold; and it
> began to be covered and uncovered from the west.

The date can be calculated as Friday, 3 June AD 1239.

In other instances, dates recorded in the past can be corrected. Clavius' *In Sphaeram Ionnis de Sacrobosco* states:

> [I]n the year 1559 about midday at Coimbra in Lusitania
> (Portugal) . . . the Moon was placed directly between
> my sight and the Sun, with the result that it covered the
> whole Sun for a considerable length of time and there
> was darkness in some manner greater than that of night.
> Neither could one see very clearly where one placed his
> foot; stars appeared in the sky, and (miraculous to
> behold) the birds fell down from the sky to the ground
> in terror of such horrid darkness.

Clavius' memory (he was writing in retrospect) cannot have served him well for the only total solar eclipse visible in Portugal between 1540 and 1600 occurred on 21 of August 1560, reaching maximum totality there just before noon.

These examples are given not only to demonstrate the precision in time and place which eclipse phenomena offer to an historical perspective, but to present something of the quality of these awesome events through the words of the writers.

Another modern insight provided by eclipses is evidence for a slowing down in the speed of the rotation of the earth. One important total solar eclipse in this connection occurred on 15 April 136 BC at Babylon. It is the only total solar eclipse yet discovered in Babylonian records and, in addition, the British Museum has two separate tablets recording the same eclipse. Calculation shows that if the earth rotated at the same speed in 136 BC as it does now, then the eclipse shadow would have passed instead over the western tip of Africa, Europe and the Baltic. This represents a time difference (measured against an 'ideal' clock) of over three hours. Some factors increase the earth's rotation, others decrease it. Gradually, however, the spin is decreasing, though variably so.

The spectacle of a rotating earth is today within reach of the direct experience of the human being. The astronauts who set foot on the moon did so during 'daytime' on that part of the moon turned to the earth. As there is no atmosphere on the moon, the sunlit sky for them was black, with stars unable to shine in the glare of day. However the earth would be visible as a large moon and, if fairly close to the sun, would appear in crescent phase and seem to rotate slowly. If the earth eclipsed the sun, the astronauts would stand in the reddish shadow of the earth.

Moon-centred astronomy experiences the earth in the way we have discussed it in relation to eclipses. The earth stands virtually motionless in the lunar, earth-facing sky, hanging perpetually in one place above the horizon and rotating on its own axis. The reason for this is that the side of the moon turned to the earth remains in this position. Sun, planets and stars will move behind the stationary earth in the course of one lunar day and night (one earth month). In that time the earth will be seen to pass through a range of phases from new to crescent, gibbous, full, etc. Figure 7.22 shows the earth full in a starlit sky as seen from that half of the moon facing the earth.

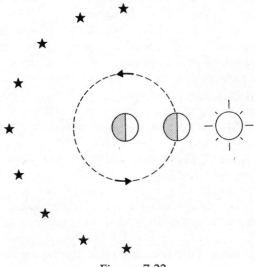

Figure 7.22

Yet the earth will not stand exactly motionless in the lunar sky. As mentioned in Chapter 6, the moon has a sideways 'dancing' movement called libration in longitude. It also makes a smaller movement up and down called libration in latitude. These two motions put together cause, for a moon inhabitant, the earth to do the same thing against the moon's horizon. The earth then 'dances' left-right, up-down in the course of a lunar day and night, tracing loops varying in shape and direction (three examples shown in Figure 7.23).

As said before, our astronauts would stand in the earth's shadow if the sun were eclipsed. People on earth, however,

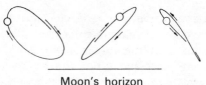

Moon's horizon
Figure 7.23

would observe an eclipse of the moon at that moment. This picture was used by Kepler, in his little book *Somnium* or 'Dream' on lunar astronomy. Here the shadow of the earth in a lunar eclipse becomes the vehicle for mysterious beings to transport a young man (a student of Tycho Brahe) along it and deposit him on the surface of the moon. From there the singular appearance of the sky is described, with an earth which 'remains fixed in place, then, as though it were attached to the heavens with a nail'. Much detail is imparted through the apprentice astronomer's eye on these strange skies. 'As the sun and stars move non-uniformly about the moon-dwellers every day' the regular rotation of features on the earth becomes their only measure of time. They cannot rely on the sun as it 'seems to make, as it were, certain jumps in relation to the fixed stars, separate jumps each day' – the sun accelerating through the zodiac at noon for those dwellers on the earthward side of the moon (a parallax effect). It is also pointed out how the phases of earth and moon are always opposite to each other at any one time – for example, when the moon is at waxing crescent as seen from earth, the earth is at waning gibbous as seen from the moon.

It is a brilliant little text and more should be known today about lunar astronomy. After all, Man has stood on the moon and we should do it the honour of understanding its sky. Yet surprisingly little is written on the subject, and the public is still left, in its films and books, with interplanetary space travellers gazing out of their craft at star-spangled skies. No one seems to have asked the real astronauts what they saw out of their portholes – a black sky with only the brightest star or planet visible. The same is the case from the surface of the day-time moon.

Back on earth, technology continues to serve the astronomer, and when a sun eclipse occurs scientists are sometimes flown in a supersonic aircraft within the moon's shadow, remaining within it for more than an hour. Returning to our back garden after voyaging along the earth's shadow with Kepler and within the moon's shadow with the scientists, we can look up at a crescent moon in early evening noticing how the unlit part glows with an 'ashen' light. When this occurs in spring with the inner crescent turned downwards like a cup filled with silvery substance, this is sometimes called the Grail Moon. After our voyage we now realise that the inner crescent is filled with our own light, for seen from the moon the earth is shining in gibbous phase, and with a bluish colour, towards it. The first person to understand this reason for the 'old moon' held in the 'young moon's arms' was Leonardo da Vinci. The ashen glow within the crescent of the moon is a signature of the earth's activity in the realm of light as it shines, normally invisibly to us, into space.

To return directly to our theme, something further should be said about eclipse periods and some particular eclipses in history. The Saros period has been mentioned, but the pecularities of the sun-earth-moon system allow several others. It has been said how unlikely it would have been for the Babylonians, particularly at the early stage of their history, to use the Saros period for predicting solar eclipses (though not lunar). The possible (approximate) eclipse periods of less than thirty years are those of the following number of months – 6, 41, 47, 88, 135, 223 and 358. The Saros is 223 months long. Babylonian records show that it is possible they used the period of forty-seven months to predict solar eclipses.

For interest, a longer cycle consists of 521 years 3 or 4 days (the number of days depending on how many leap years there are in that time) which has eclipses repeating near a particular latitude. For example, the solar eclipse of 17 June 157 BC was total in England; that of 16 June AD 364 was total in Scotland; that of 16 June 855 was total in Scotland; that of 16 June 1406 was total in Belgium; and that of 29 June 1927 was total in Wales, Lancashire and Yorkshire. A cycle which, like the Saros, restores the diameters and motions of the sun and moon almost exactly is that of 1,805 years and a few days. For example, the eclipse of 1927 (mentioned above) occurred in the morning in Scotland, and its partner on 21 June 122 was total in the Shetlands area in the evening.

Then there is the Metonic cycle of eclipses which produces an eclipse every nineteen years on the same calendar date.

However, this is not very useful for prediction as the cycle only contains four or five eclipses and it would be difficult to know when the series began or ended. The nineteen-year cycle essentially relates to the calendar, for after 19 years the phases of the moon recur on the same days of the same months, within about two hours. This cycle was used by the Greeks to predict the days on which their religious festivals, determined by the moon's phases, should be celebrated. It is still used by churches today and the number of a year within a particular span of nineteen years is called the Golden Number. The Christian calendar dates the start of its Metonic series from the year 1 BC. It is said that the Greek astronomer Meton (fifth century BC) had the years of this nineteen-year cycle inscribed in gold letters on a temple in Athens, hence the name Golden Number.

Some further details can be given of particular eclipse phenomena. At the time of writing, the United States of America and Canada have experienced their last total solar eclipse this century – on 26 February 1979. The next to appear there will be on 21 August 2017. But in Britain the track of a total solar eclipse will pass over Cornwall on the morning of 11 August 1999, and continue eastwards to pass about 35 kilometres north of the centre of Paris (near Chantilly). The track will begin in the Western Atlantic, cross Europe, the Black Sea, Turkey, Iran and India, and end in the Bay of Bengal. This also is the year and season mentioned in a prophecy by Nostradamus – 'In the year 1999, and seven months, from the sky will come the great King of Terror. He will bring back to life the great king of the Mongols. Before and after War reigns happily.' The 1999 eclipse belongs to a Saros series which began in 1639 at the north pole. The series first became total in 1891 and the eclipse of 29 June 1927, visible in England, was also a member of it.

Lunar eclipses suffer less distinction historically as they do not relate to specific locations but rather to about half of the earth's surface at once. However, the dating can be of interest, and one lunar eclipse that may be considered worthy of note is that of 3 April AD 33. This is considered by some authors to be the day of Christ's crucifixion (see Ormond Edwards's *A New Chronology of the Gospels*). On that day the moon was eclipsed partially to a maximum of 60 per cent of its disc one hour before moonrise (about 6.15 p.m.) at Jerusalem. When the moon rose at sunset it was still eclipsed (by 20 per cent) and remained partially in the earth's umbra for half an hour afterwards. This, plus its low position in the atmosphere, would have given the moon a red appearance. This phenomenon is pointed to by Colin Humphreys and W. G. Waddington (in the science journal *Nature* in December, 1983) as convincing evidence that the Crucifixion indeed took place on Friday, 3 April AD 33. Among other sources they cite *The Apocryphal New Testament* which has a 'report of Pilate' saying that at the Crucifixion 'the Sun was darkened; the stars appeared and in all the world people lighted lamps from the sixth hour till evening; the Moon appeared like blood'. This eclipse was one of a Saros series which began as a penumbral eclipse at the moon's descending node (dragon's tail) on 4 June 473 BC and ended on 11 July AD 808. The series produced total eclipses from 16 May AD 105 to 1 September AD 285.

However, the sun was said to go dark on the afternoon of the Crucifixion, and of this Sacrobosco wrote in his *Sphere* –

[W]hen the sun was eclipsed during the Passion and the same Passion occurred at full moon, that eclipse was not natural – nay, it was miraculous and contrary to nature, since a solar eclipse ought to occur at new moon or thereabouts. On which account Dionysius the Areopagite is reported to have said during the same Passion, 'Either the God of nature suffers, or the mechanism of the universe is dissolved.'

Also, reports of a darkening of the sun for a three-hour period at the Crucifixion do not tally with a solar eclipse, which can only last a few minutes.

Examining lunar eclipse frequency for a moment, we find that the maximum number of lunar eclipses (including penumbral) which can occur in a year is five. This last happened in 1879 and will take place again in 2132. In one calendar year there can be a maximum of three total lunar eclipses and this has occurred twice this century, in 1917 and 1982. There are long gaps prior to and following these events, so that the 'triple' before 1917 was in 1544 and the next 'triple' after 1982 will be in 2485. In one century an average of about 241 lunar eclipses (including penumbral) takes place. In the period 1501 to 2200, the twenty-first century has the highest number of total lunar eclipses (eighty-four) with the twentieth century ranking second with eighty-one. The nineteenth century had only sixty-two. An additional phenomenon is that four total lunar eclipses can succeed each other at intervals of six months. This is referred to as a 'tetrad' and several of these take place in groups of years which are separated by an average period of 586 years. This interval was discovered by the Italian astronomer Schiaparelli (1835–1910). No tetrad took place between 1582 and 1908, while from 1909 to 2156 there will be a total of sixteen tetrads, one occurring in 1985–6.

To return to the Saros, it would take a very long period of time for this cycle no longer to work. So far it has accompanied humanity's development over many thousands of years, threading it through with amazingly regular rhythms and calling on Man's pure thinking to unravel and understand the thread with mathematical insight. If someone asks 'What is an eclipse?', the answer is – 'When a dragon comes along and swallows up the sun or moon.' The dragon today is mathematics, but it is a benign creature, leading us into the clear air of number, plane, point and line which stands behind the phenomenon of darkening.

The study of eclipses is not diminished by an approach based on quantitative calculation and celestial mechanics, for the very details of their number and geometry are remarkable enough in themselves and speak their own language of order and organisation in the relationship between sun, moon and earth. In ancient times the priest-astronomer felt that the advent of a particular eclipse carried influence and meaning with it. Today, science does not recognise this aspect, but nevertheless can produce its own experiences out of the phenomena to match it. For example, experiments have been conducted in Switzerland on the behaviour of plant saps in connection with the moon, as mentioned in the last chapter. It was found that during a total eclipse of the sun in New Zealand and the southern Pacific Ocean, plant saps in Switzerland were weakened in their activity. This suggests that the earth, like a living body, is a whole organism in itself and what happens in one part affects the rest. Science has further steps to make in understanding such aspects of nature. The phenomenon of eclipses can yield as much awe and insight today as it did in ancient times, though in a new way.

Chapter 8
The Interior Planets

The word 'planet' comes from the Greek meaning a wanderer. Sun and moon were seen to wander against the background of the stars, the sun moving through all the zodiacal constellations in a year, the moon in a month. This was a 'direct' motion eastwards. Therefore sun and moon were designated as planets along with five other wanderers – Mercury, Venus, Mars, Jupiter and Saturn.

The average daily speed of their motion varies, the moon being the fastest and Saturn the slowest (Figure 8.1). This was

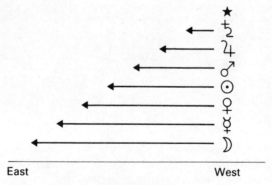

East   West

Figure 8.1

the ancient order of the planets, experienced as movement in time. If we now place these planets in this sequence round a circle, we obtain the sequence of the days of the week (Figure 8.2a). In English, Saturday, Sunday and Monday retain their obvious derivation from Saturn, sun and moon. In French the planetary connection is evident in mardi (Mars' day), mercredi (Mercury), jeudi (Jupiter) and vendredi (Venus). One obtains

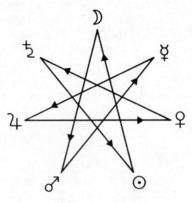

Figure 8.2a

the same weekly sequence if one designates a planet to every hour of the day (as was done anciently) in the order in Figure 8.1, from Saturn to moon, and names each day according to the planet occupying the first hour (Figure 8.2b).

Hours ——→

Days
↓

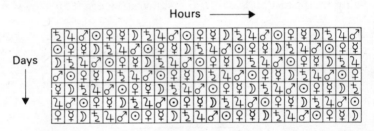

Figure 8.2b

In this time-sequence of planetary order, Mercury and Venus lie between earth and sun, and we shall refer to them as the interior planets. (An older designation for them has been 'inferior planets'. In this text it is preferred to employ the term 'interior', also used by the astronomer V. A. Firsoff. 'Inner planets' has been applied to them by some writers, though the modern astronomer means this to denote celestial bodies (including the earth) within the asteroid belt.) A clear distinction has to be made, as far as earth observation is concerned, between these interior planets and the 'exterior' ones, Mars, Jupiter and Saturn. The phenomena of appearance of each type are different, and by nature they fall into two separate groups.

Firstly, the interior planets. If one took the light of all the naked-eye visible stars in a hemisphere and brought it together in one point, it would have the same magnitude of light as Venus at its brightest. Venus is always far brighter than any star or, in fact, than any of its planetary companions. Indeed, it can be seen in daylight if looked for carefully at the right time and can cast a shadow at night. Its partner Mercury, on the other hand, is modest in brightness by comparison and was anciently referred to as the 'faint' planet. Mercury and Venus are very different, even opposite, in individual character but they perform the same interior planet dance witnessed from earth.

The nature of their performance is that essentially they make their appearance as morning or evening stars. They appear in the vicinity of the sun for a while after it has set or before it rises, their territory being easterly or westerly above the horizon. While doing this they describe curves in relation to this horizon which vary in shape from one appearance to another – Venus in particular displaying variations. Venus makes long, elegant forms over a comparatively extended period of time, while Mercury makes quick, frequent curves of short duration. Venus proudly announces itself for all to see; Mercury slips quietly into the sky and challenges one's knowledge and eyesight to find it.

For example, within a six-month period, Venus can make the movement shown in Figure 8.3, but Mercury in that time is capable of appearing three times – twice in the east and once in the west, as in the case of Figure 8.4.

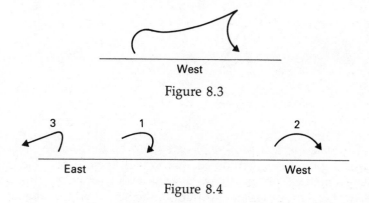

West

Figure 8.3

East   West

Figure 8.4

The geometrical background to the interior planets is that, seen from earth, Venus and Mercury appear to oscillate from one side of the sun to the other. Their movements in relation to the sun can be deduced to be as in Figure 8.5. When either

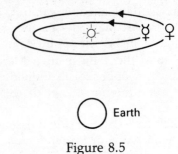

Earth

Figure 8.5

planet is close to the sun for an observer on the earth, it is not visible. But when they draw to left or right of the sun, they can be seen in the evening or morning sky (Figure 8.6). The planet then moves in relation to the sun and the sun and ecliptic move in relation to the horizon from day to day (Figure 8.7), which combines to give the sort of curve shown in figure 8.3.

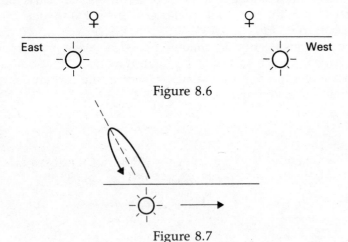

Figure 8.6

Figure 8.7

The speed at which these planets move on their curves, and their brightness as they do so, also varies. Seen in plan view, the geometrical picture shows that the widest angle which the planets can move to left and right of the sun is where tangent lines from the earth touch the planets' orbits (Figure 8.8). The tangent points are not opposite each other on their separate circles. Therefore, seen side-on, as the situation is for the earth observer, an interior planet will describe an ellipse, but the points V and V' do not represent the ends of a diameter through the centre of the orbit (Figure 8.9). In space, the planet has further to go between V' and V than between V and V'.

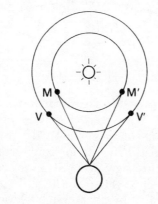

Figure 8.8

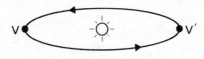

Figure 8.9

The result is that in travelling from V' to V the planet will appear to move more slowly than from V to V'. At the tangent points themselves it will appear to stop its movement east or west in relation to the sun, and turn (Figure 8.10).

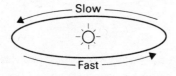

Figure 8.10

When the planet is between earth and sun or directly on the far side of the sun, it is invisible in the sun's light. These positions are, respectively, inferior conjunction and superior conjunction (conjunction symbolised by ☌). At the tangent points it is at greatest elongation (Figure 8.11).

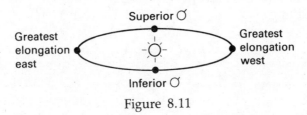

Figure 8.11

An interior planet is thus fastest in relation to the sun at inferior conjunction. With respect to brightness, Venus and Mercury do opposite things. Venus has two positions of 'greatest brilliancy' between inferior conjunction and greatest elongation, but Mercury's greatest brilliancies occur, in general, near superior conjunction (Figures 8.12 and 8.13). The reasons for this are discoverable by telescopic observation, which will be discussed in Chapter 12.

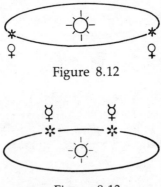

Figure 8.12

Figure 8.13

There follow from the above certain laws of movement and light concerning both planets. Venus will move swiftly into the morning sky, brightening quickly before dimming and leaving the sky slowly (Figure 8.14). Mercury will move swiftly but

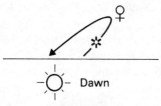

Figure 8.14

dimly into the morning sky, brightening as it departs slowly (Figure 8.15). On the other hand, Venus will move slowly but dimly into the evening sky, departing from it swiftly and brightly, while Mercury will move slowly and brightly into the evening sky, leaving it quickly and dimly.

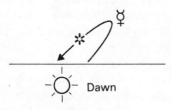

Figure 8.15

Venus and Mercury are guardians of eastern and western horizons which are the entrance and exit places for the stars. Particularly in the case of Venus, one is aware of its strong presence when presiding over, say, the evening western sky – and when it is not there one feels the loss and the character of the sky is different. One is made conscious of when the queen is in her palace or when she is not. When in her western home she sets serenely in the darkness and at such times was a goddess of love for the Babylonians. When a morning star Venus rises into daylight, battling to outshine even the sun, and at such times it was known to the Babylonians as the goddess of war. One tablet identifies Venus as female at its sunset appearance and male at sunrise. The morning appearance of Venus was also connected in ancient times with Lucifer as a herald of the light, or Phosphoros.

Mercury, on the other hand, complements Venus with its short but rhythmic visits low over eastern and western horizons. In this role it keeps close to the earth and seems to mediate at the threshold between above and below. It was known as the messenger of the gods, the Greeks naming it Hermes at its evening appearance and Apollo in the morning. Its light is entirely different to that of Venus. The latter announces itself and radiates towards one. Mercury asks the observer to be active and look with concentration. When it is seen, one is often met with a sharp, steady point which pierces the veil of light shed by the nearby sunset or sunrise. One Turkish name for it was Tir, meaning an arrow. Mercury is, in fact, a bright planet, but is mostly seen when the twilight glow of the sun is in the sky, and so appears relatively faint. Proclus Diadochus (410–485 BC), head of the Platonic Academy in Athens, described it as 'Mercury the elusive' and the Platonic philosopher Apuleius (born c. AD 124) called it 'the nimble one'. It seems always to shine from afar, from some distant realm. It can stand unnoticed and one has to seek it out. To see Mercury consciously for the first time is a moment of initiation into the lore of the stars.

Quicksilver has for centuries been associated with the planet Mercury, and researcher Agnes Fyfe in Switzerland has found, in experiments with plant sap and metal salts, a response in quicksilver, in particular, to movements of the planet Mercury. She also found a connection of the sun with gold, the moon with silver, and Venus with copper. Medieval alchemists made the same connection between these particular metals and planets, further relating iron to Mars, tin to Jupiter and lead to Saturn – a correspondence supported in experiments using metal solutions and capillary action by L. Kolisko in Switzerland.

Venus takes over nineteen months (584 days) to swing from one side of the sun to the other and back again, as seen from earth. This is its synodic period and means that any of its positions in relation to the sun (e.g. greatest evening brilliancy) is repeated in that time. But the time it takes to move from western elongation to eastern is greater than from eastern to western elongations, as the latter journey is swifter (see Figure 8.10). The passage from furthest west of the sun, through superior conjunction to furthest east, takes almost fifteen months, while the opposite passage through inferior conjunction approaches five months only.

Mercury's synodic period is four months (about 116 days on average). Its journey through superior conjunction from western elongation to eastern is 2½ months, while the journey from furthest east to furthest west takes 1½ months.

Venus moves far enough away from the sun to be seen in darkness against a background of stars. Taking a particular point in its synodic revolution, say greatest elongation east, it may happen that this places it near the star Hamal in the Ram. Nineteen months later, greatest elongation east will again take place, this time near the star Antares in the Scorpion. These events are almost exactly seven equal constellations apart and the result is that in eight years the eastern elongation point will return close to where it started – only about 2 degrees short – marking out a pentagram on the zodiac (Figure 8.16).

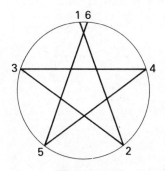

Figure 8.16

If such observations were made for Mercury under good conditions, say near the earth's equator, then greatest elongations east would step almost round the zodiac in twelve months (though not in a calendar year) falling short by about half a constellation. There would be four eastern elongations in that time, forming an unfinished triangle (Figure 8.17). Greatest elongations west would form a similar triangular figure close to the other one, differing by only a few degrees – as, for example, in Figure 8.18 (which covers a period of about fourteen months). The synodic positions of Mercury seen against the stars repeat themselves with accuracy after forty-six years.

The best times of the year in the northern hemisphere for observing elongations of the interior planets are in spring (when the ecliptic has a steep angle to the horizon at sunset) and autumn (steep angle at sunrise). However, it so happens that Mercury is in that part of its motion which is nearest to the

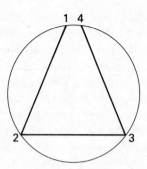

Figure 8.17

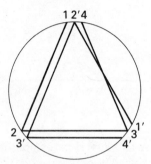

Figure 8.18

sun (perihelion) when it is spring and autumn in the northern hemisphere, therefore at these times of 'best seeing' the planet is in its least favourable position for elongations and will be placed at an angular distance from the sun of only 18 or so degrees. In the southern hemisphere Mercury is best seen and is at its *furthest* from the sun (about 28 degrees) in spring and autumn, so its visibility is much increased.

Mercury makes an appearance in the morning sky every four months, and in the evening sky every four months. Thus, in a calendar year, six completed appearances fall between January and December, diagrammatically shown in Figure 8.19. Perhaps one may be permitted to note that these frequent passages from one side of the sky to the other are reminiscent of the caduceus traditionally carried by the god Mercury – the staff with two snakes intertwined round it (Figure 8.20).

As Mercury and Venus pass between sun and earth, it can be expected that from time to time they will stand directly

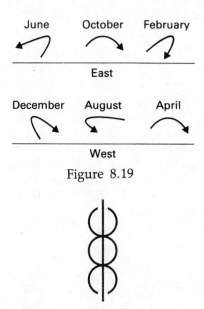

June   October   February

East

December   August   April

West

Figure 8.19

Figure 8.20

between and appear as black dots on the face of the sun. The twelfth-century Arab astronomer Alpetragius, noting that Mercury had never been seen to cross the sun's face, concluded that the planet was self-luminous or translucent. But he did not realise that Mercury is too small to be seen thus by the naked eye. Mention of Mercury's 'transits' of the sun's disc will be delayed until Chapter 12 and a discussion of the telescope, the latter being necessary to observe the phenomenon.

However, Venus is large enough to be seen in transit with the naked eye – though it must be done using protection against glare or indirectly with a camera obscura, as in the case of solar eclipses. It is safest, to avoid retinal damage, for the sun never to be observed directly but always using the camera obscura method of projecting the sun's image through an aperture onto a screen or wall.

Transits of Venus are rare, there being none this century. The last was in 1882 and the next will be in 2004. The first historical prediction of a transit of Venus was calculated by Kepler for 1631. An attempt to observe it was made by Father Pierre Gassendi who, however, saw nothing. But Kepler's prediction was right and the transit took place, though after sunset in France. Yet Kepler did fail to predict a second transit for eight years later and this was rectified by a young English clergyman, Jeremiah Horrocks of Lancashire, who completed his calculations just before the event on 4 December 1639. This turned out to be a Sunday (not good due to ecclesiastical duties) with bad weather (not good for astronomical observation). However, on returning from church to his darkened room into which an image of the sun was projected through a telescope, a break in the clouds allowed him to see the dark spot of Venus already advanced upon the solar disc and he recorded the event in a report. Figure 8.21 shows the path of the transit he witnessed which would have lasted 6½ hours.

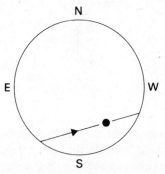

Figure 8.21

Transits of Venus occur only in December or June in periods of 113½ years plus or minus eight years. For example, the transit of 1639 is followed by others in June 1761, June 1769, December 1874, December 1882 and June 2004. They succeed each other at intervals of 121½ years, 8 years, 105½ years, 8 years, 121½ years . . . and so on. A series of five transits covers a period of 243 years, after which the combination of year intervals is repeated for another 243 years.

The transit of 2004 will take place on 8 June and will be wholly visible from Britain, while the following one eight years later on 5–6 June 2012 will start before the sun rises in Britain and only the end will be visible from there. Figure 8.22 shows these two events, the first of which will last for over five hours and the second for more than six. Transits are miniature solar eclipses with the disc of Venus only about one-thirtieth of the size of the sun's and moving across it in the opposite direction

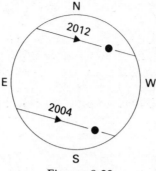

Figure 8.22

to the moon.

It is sometimes said that people with extremely acute vision can see Venus develop phases like the moon. The mathematician Gauss is said to have tried to surprise his mother by asking her to look at Venus through his telescope, but the crescent shape she saw was already familiar to her and she simply asked why the crescent was turned the wrong way (the astronomical telescope inverts the image). The surprise was Gauss's that his mother could see the phases of Venus with the naked eye. I imagine that the eyesight of few readers would be equal to that told of Gauss's mother, and shall reserve the phase appearances of the interior planets until discussion of the telescope in Chapter 12.

The interior planets are 'wandering' rebels against the celestial sphere of stars, displaying independent action. Their centre, or central reference point, is the sun, to which they keep close. Also independent of the stars are the exterior planets, and the sum of planets (even including the moon) relates itself by nature to the rhythms of their chief wanderer the sun as it moves through its own self-created zodiac which begins at the spring equinox point (see Chapter 5). One can treat the star sphere as fixed and regular, and the planets as the element of animated life with rhythms of its own – waxing and waning, lifting and falling, appearing and disappearing. These rhythms can be found reflected in many aspects of life on earth. Blended with them is the turning of the star sphere which provides a slow, majestic background. In this context can the script of the planets be read and their appearances become consciously part of our environment.

Chapter 9
The Exterior Planets

Just as the essential characteristic of the interior planets is to describe curves in relation to the horizon, so the main characteristic of the exterior planets is to describe curves or loops in relation to the stars. The interior planets stage their performance at the horizon at sunset and sunrise, whereas the exterior planets can appear all night, visibly traversing the sky from east to west and shining boldly in the south at midnight. This means that they stand opposed to the sun at certain times, instead of always being drawn back into it before venturing too far.

An exterior planet moves round the zodiac in an anti-clockwise direction in the course of the year, but slower than the sun (see Figure 8.1). Therefore the sun catches up with and overtakes it periodically. At the moments of being overtaken it is, of course, in conjunction – 'superior' conjunction, as the planet is beyond the sun. At another stage the planet is in 'opposition' (symbolised by $\mathcal{S}$) to the sun when the planet appears on one side of the zodiac and the sun directly opposite on the other. The earth-centred picture is shown in Figures 9.1 and 9.2.

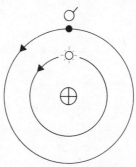

Figure 9.1

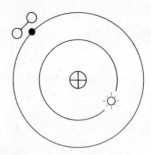

Figure 9.2

Mars, Jupiter and Saturn always enter the evening sky in the east, Mercury and Venus always enter it in the west. Technically, an exterior planet begins its life as an evening star when it is in opposition and therefore rises at sunset. As the sun is moving about 1 degree eastwards every day through the zodiac to catch it up, the exterior planet will rise earlier each evening. This will place it higher in the sky from evening to evening at sunset (Figure 9.3). Its appearance as an evening star ends when it sinks into the sunset glow when the sun has caught it up (Figure 9.4) and it moves into conjunction.

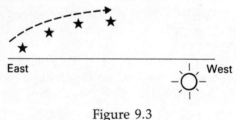

Figure 9.3

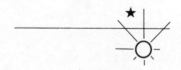

Figure 9.4

When the exterior planet moves into the sky at opposition it is at the other side of the horizon to the sun and away from its light. For the rest of the evening it climbs into a dark sky with the stars as background, and it is in this environment that its special movements and variations of brightness display themselves. Interior planets never appear in opposition to the sun and therefore cannot emulate the exterior planets' movements against the stars.

If we begin observing an exterior planet earlier than opposition, say at quadrature (90 degrees from the sun), when it rises at midnight (Figure 9.5) to when it sets at midnight at the other quadrature position (Figure 9.6) then we have opposition

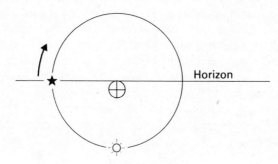

Figure 9.5

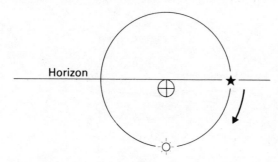

Figure 9.6

between these times and a chance to obtain an overview of an important part of the planet's motion. Seen against the stars, over a period of months, it will perform a loop or similar form (Figure 9.7). Often the modern person cannot believe at first that exterior planets make such forms before our eyes, until he is shown a heliocentric reason for it. Then he has the rational, geometrical explanation, but the sense of wonder and the qualitative feeling for the appearance is lost. Mechanical laws here overtake the primal experience.

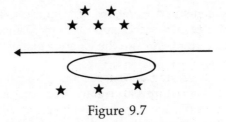

Figure 9.7

The phenomenon is that a loop, or similar, is made during which the planet speeds up while moving backwards from left to right (retrograde) and when at or near the middle of this backward motion it shines brighter than at any other time (Figure 9.8). At this brightest stage it is retrograding swiftly westwards and is in opposition to the sun precisely in the middle of the loop. Therefore half of the movement will take place when it is technically an evening star and half when it is a morning star (Figure 9.9 shows the two halves). As evening star it rises before sunset, as morning star it rises before sunrise.

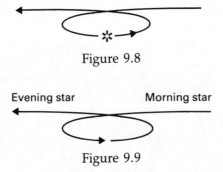

Figure 9.8

Evening star   Morning star

Figure 9.9

It can be seen that, in relation to an east-west direction along the zodiac, an exterior planet reaches two 'stationary points' against the background of stars (points S^1 and S^2 in Figure 9.10). At these positions the planet moves as the stars move, rising at the same speed and setting at the same speed. If we mentally project the loop onto the eastern sky at sunset, then we see the following consequence. At opposition, the retrograde part of the planet's loop movement will carry it swifter than the stars into the evening sky from night to night (Figure 9.11). At the stationary point S^2 (Figure 9.10) it will coincide with the speed of the rising stars, and after that it will rise slower than the

Figure 9.10

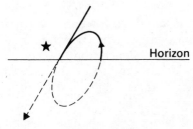

Figure 9.11

stars. If we then follow the planet to superior conjunction, we find the sun overtaking it, both moving westwards among the stars (Figure 9.12). Therefore, before superior conjunction the planet will set in the evening western sky more slowly than the stars (Figure 9.13). After superior conjunction it will rise more slowly than the stars into the morning eastern sky (Figure 9.14). So the exterior planet rises and sets swiftly and brightly in the middle of its loop at opposition, and rises and sets reluctantly and dimly when close to the sun near conjunction. Such are the exterior planets' relation to the stars and, consequently, the horizon – quite different to that of the interior planets.

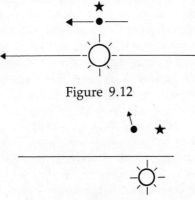

Figure 9.12

Figure 9.13

Figure 9.14

Interior planets are bright when they are near the sun, exterior planets are bright when separate from it. The exterior planets achieve independence from the sun and, at that time, shine brightest – as does the moon. But in addition they move back against their normal motion eastwards, intensifying their activity and the gesture of independence. They do this between quadrature, opposition and the next quadrature, while the interior planets never reach quadrature and perform their piece on the other side of the circle of configurations – in the region where conjunctions take place. Exterior planets, on the other hand, never have inferior conjunctions; along with the stars of the zodiac, they pass through opposition instead. This is an indication of their closer affiliation to the stars.

We can now turn to individual planets, taking Mars first. There are wide variations in the brightness of Mars' red light. When near to the sun it shines as modestly as the Pole Star, but when in opposition to the sun it can exceed the brilliance of Sirius. It also has variations in its level of greatest brilliancy, with a peak of brightness when its loop takes place among the stars of the Waterman. Then Mars is at its very closest to the earth. This peak of greatest brilliancy takes place about every fifteen years and is at its best if opposition occurs on about 28 August. The oppositions of 1986 and 1988 both take place close to the favourable position in the Waterman, the former on 10 July and the latter (slightly brighter) on 28 September.

In fifteen years the loops of Mars move round the zodiac in seven steps, returning to the area where the first loop was made with the seventh loop. Figure 9.15 shows opposition points of Mars from one favourable position (1971) to another fifteen years later (1986). Each step or interval between loops represents a synodic period of Mars, lasting an average of two years and seven weeks. Therefore the planet performs a loop during alternate years. This is the longest synodic period of all the planets. Figure 9.16 shows the distance along the zodiac

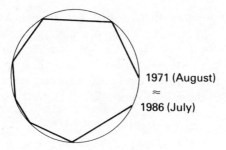

1971 (August)
≈
1986 (July)

Figure 9.15

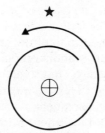

Figure 9.16

which it travels between oppositions. The zodiac phenomena of Mars repeat themselves accurately every seventy-nine years during which there are thirty-seven loops.

Like the other exterior planets, Mars performs four basic types of loop, with minor variations. When above the ecliptic the loop is turned upwards, when below it is turned downwards, and when it crosses the ecliptic from above or below during opposition, it assumes a zig-zag shape (Figure 9.17). The zig-zag is

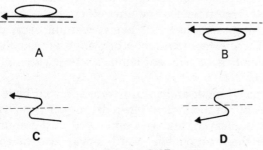

Figure 9.17

the least frequent of these forms and occurs, in the case of Mars, when opposition takes place among the stars (not the sign) of the Ram in November, and among the stars of the Scales in May. In these areas lie the ascending and descending nodes of the planet respectively. Conjunction with the sun takes place between the loops (Figure 9.18). Here the planet is at its dimmest.

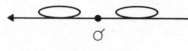

Figure 9.18

Mars makes the largest loop of the exterior planets. Its retrograde motion between stationary points covers over half a zodiac constellation which, however, it achieves quickly – in about 2 months and 12 days – compared with the retrograde motions of other exterior planets.

There is strength and swiftness in the apparent movements of Mars. It waits longest before entering the midnight sky to perform, then does so in large measure and with fiery colour. The Chaldeans named it Negral – the king of conflicts and master of battles. In Persia it was known as Bahram and Pahlavani Siphir, or the celestial warrior. The Greeks called it Ares, the god of war. The god Mars was the legendary father of Romulus, founder of Rome, and our boisterious month of March is named after the same god. There is a story that Mars and Venus had a secret love for each other and met only during the night when Apollo, the sun god, could not see them. Mars's servant, Alectryon, was appointed to keep watch for the sunrise but on one occasion fell asleep at his duty and the lovers were discovered and ridiculed by the other gods. Mars angrily changed Alectryon into a cock to give warning every day of the sunrise. These exterior and interior planets are like Romeo and Juliet – members of different families whose close relationship is irregular.

The names Jupiter and Saturn were assigned high positions in mythology. In the Indian Rig-Veda, Jupiter was Dyaus Pitar, or Father of Heaven, and was embodied in the entire sky. In Babylonia the creation god Marduk was also called Nebiru, usually meaning the planet Jupiter. Nebiru can also mean the whole central band of the heavens, or a central point or pole. The planet was known by the Babylonians as the 'bull of light' and as the shepherd of the stars.

Deities took on different planetary and stellar appellations at various times. For example, the goddess Ishtar was the star Capella in January and February, but in May and June the god Marduk was Capella. Marduk could also be the daily sun, Jupiter, Mercury or the star Regulus, depending on the season. Gods assumed their sky appearances according to the time and nature of the phenomena, which they wore like celestial raiment. They could also take on more than one appearance at the same time.

Saturn has enjoyed an exalted position among the planets. For the Babylonians it could even replace the sun, when the latter disappeared below the horizon, and was known as 'sun of the night'. The Greek historian Diodorus (c. 50 BC) said of the Chaldeans – 'the star which the Greeks name Kronos (Saturn) they call the "star of the sun" because it is the most prominent and gives the most numerous and most important predictions.'

Jupiter and Saturn have the least irregular or retrograde motion of the naked-eye planets, and come closer to the sun's regular movement against the stars. Their loops are relatively small – Saturn's being about half the length of that of Mars, with Jupiter's between the two in size. For Saturn the zodiacal distance travelled between the extremities of two loops is about the same as the length of the loop itself, while for Jupiter the distance is about twice one of its loops. Mars, as indicated, travels more than one circuit of the zodiac between one loop and the next.

Unlike Mars, Jupiter and Saturn perform loops with a duration of just over a year between them. Saturn's synodic period is about 1 year and 13 days, and Jupiter's 1 year and 34 days. With its larger loops and longer distances between them, Jupiter moves round the whole zodiac more quickly than Saturn, taking almost twelve years and producing almost eleven loops in that time, whereas Saturn moves through the zodiac in about 29 years, completing just over 28 loops. The loops of Jupiter and Saturn are smaller, flatter, and are formed more slowly than those of Mars. Their flatness makes it difficult to distinguish the zig-zag shapes which occur twice in their circuits round the zodiac.

Jupiter in opposition is usually the brightest of the exterior planets (only Mars can occasionally surpass it). At best it can out-shine Sirius, and at its least brilliant it appears about the same as that star. When oppositions occur in October among the stars of the Fishes or Ram, the planet achieves its highest level of brightness.

Saturn, on the other hand, has two periods of brightest oppositions, occurring in winter in the stars of the Bull or Twins and in summer in the Scorpion or Archer. When most bright it shines more strongly than Arcturus, and this takes place during the winter oppositions. The planet's two brightest points are at opposite sides of the zodiac and occur every fifteen years, the reason for these brightenings not being known until the development of the telescope and observations by Christiaan Huygens in Holland in the seventeenth century (see Chapter 12). Figure 9.19 shows those positions in the zodiac constellations where Jupiter and Saturn are brightest.

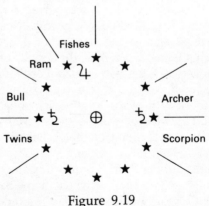

Figure 9.19

Every twenty years Jupiter and Saturn stand in conjunction with each other. There are two important but quite different ways of determining this moment – either in relation to the ecliptic or to the celestial equator. The date of a conjunction can vary by days or weeks, depending on which of these two approaches is used. Figure 9.20 illustrates the reason – the planets must lie on a line at right angles to the ecliptic or the

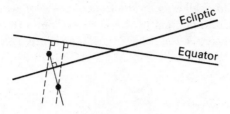

Figure 9.20

equator. The figure shows a conjunction by the ecliptic method, but not by the equatorial. The ecliptic relationship with the planets seems the more natural one for observation, and it often marks almost exactly the planets' smallest angular separation or closest approach.

Conjunctions of Jupiter and Saturn are traditionally called Great Conjunctions and were looked upon with special interest in the days when astrology was part of science. The sixteen-year-old Tycho Brahe observed the Great Conjunction of August 1563 with the eyes of an astronomer as well and, holding the end of a pair of geometry compasses close to his eye, measured the planets' angular separation. He found that the forecast in the *Ephemerides* of Johannes Stadius, based on the work of Copernicus, was several days in error, and was then to devote his life to establishing accurate astronomical positioning. As Great Conjunctions take place approximately 117 degrees apart, they step through the zodiac in triangles. Zodiacal signs in triangular relationship were referred to as 'trigons' by the astrologer and shared the same element of earth, or air, or fire or water. The Great Conjunction of April 1583 took place within the sign (not the star constellation) of Pisces, marking the conclusion of the domination of the 'watery' signs which had been the zodiacal areas for Great Conjunctions since 1365. These conjunctions remain in a particular trigon for about 200 years and pass through all four trigons or signs of the zodiac in about 800 years. This latter cycle was considered of much importance in earlier times, Tycho Brahe referring to its relevance to history in his writings on the New Star of 1572.

He spoke of the 800-year period beginning with a Great Conjunction in the 'fiery' sign of Sagittarius in December 1603 as ushering in an age such as foretold by the prophets Isaiah and Micah when the lion shall eat straw like the ox, and the suckling child shall play on the hole of the asp (see also Chapter 10). Great Conjunctions in the zodiacal signs this century are shown in Figure 9.21 (1980/1 marking an initial entry into the air signs).

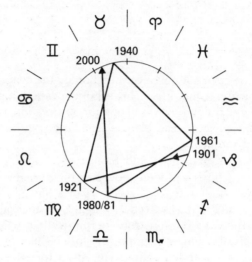

Figure 9.21

There are occasions when Jupiter and Saturn are in opposition less than 1.7 days of each other, and then a 'triple conjunction is likely. The planets' loops coincide in longitude and Jupiter, with its bigger loop and faster movement, passes in line with Saturn three times (Figure 9.22). These events are rare,

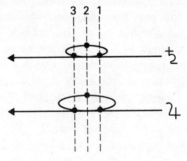

Figure 9.22

happening on average about every 139 years. The last one took place in 1980/1 and the next will be in 2238/9 (calculated in relation to the ecliptic). If we include all the exterior planets in the possibilities of triple conjunctions, then for Mars and Jupiter the last was in 1979/80 and the next will be in 2123; and for Mars and Saturn the last was in 1945/6 and the next will be in 2148/9. Triple conjunctions can also take place between exterior planets and stars, and those occurring with bright zodiac stars towards and around the end of this century are: Saturn/Antares 1986; Mars/Aldebaran 1990/1; Mars/Regulus 1994/5; Jupiter/Antares 1995; Jupiter/Aldebaran 2000/1; and Saturn/Aldebaran 2001/2. It is interesting to note that a triple conjunction of Mars and Spica is always followed two years later by a triple conjunction of Mars and Antares.

Ptolemy gave planetary qualities to stars, stating that Aldebaran had the nature of Mars; Regulus that of Mars and Jupiter, and Antares that of Mars and to a moderate extent Jupiter.

In triple conjunctions between planets, these celestial bodies come into closest relationship to each other while at the same time being at their strongest and brightest stages in opposition to the sun. They virtually share an opposition point in the zodiac and it is as if an invisible knot then tied them together.

For a moment we can turn to a unique aspect in the study of planetary motions in the system of the Platonic mathematician Eudoxus (c. 370 BC). This brings us again to the figure of the lemniscate which can, under certain circumstances, have special connections with movements in longitude and latitude – with movements on the celestial sphere related to the primal orientations of left-right and up-down. Eudoxus tried to incorporate the gesture of planetary loops into a geometrical model which contained built-in lemniscates. He devised a system of concentric spheres rotating within each other. In a group of spheres related to a particular planet, the pole of one sphere was attached to the inner surface of the other. Four spheres were assigned to each of the five planets, the planet itself being attached to the innermost of the four spheres. The two outermost spheres provided the planet's diurnal motion and the movement along the zodiac. The two innermost ones moved the planet on an apparent lemniscate along the ecliptic, superior or inferior conjunction with the sun being at the double point (Figure 9.23). The combination of these movements pro-

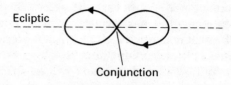

Figure 9.23

duced an apparent motion among the stars. Examples of possible resulting motions are shown in Figure 9.24. This was an attempt to give a generalised, characteristic form to planetary behaviour and did not exactly reproduce the observed phenomena. The greatest discrepancies were in the cases of Mars and Venus, neither of which could be made to display retrograde motion. The lemniscate curve was called a hippopede by the Greeks because horses were cantered on this form during practice in the riding school.

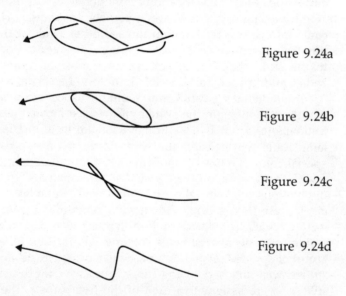

Figure 9.24a

Figure 9.24b

Figure 9.24c

Figure 9.24d

Altogether, Eudoxus' system contained twenty-seven concentric spheres (three each for the sun and moon and one for the stars). The system was later improved with additional spheres (bringing the total to thirty-three) by the astronomer Callippus of Cyzicus. For this he worked with Aristotle in Athens. Additional spheres were supplied not only to the planets

Venus, Mercury and Mars, but to the sun and moon also. Sun and moon were given five spheres each which appear to have incorporated lemniscate motions into the progress of these luminaries along the ecliptic, accounting for unequal motion in longitude. Aristotle himself then added several intermediary spheres between the planets, designed to unite their motions into one mechanical whole with moving contact throughout, replacing Eudoxus' purely abstract, geometrical system in which each planet had an independent working. A further step was thus taken towards 'saving the appearances'.

Mars has now been reached by unmanned spacecraft which have analysed soil and taken 'photographs' from remote control on earth. This planet would certainly be the next goal for Man to set foot on beyond the moon. We are now familiar with pictures of the red, dry Martian soil and pink skies seen through the dust. What, then, is the Martian astronomy?

Measured against earth time, the Martian year is almost twice as long as the earth's (687 earth days) and the solar day is longer, being 24 hours 37 minutes in earth time. There are 668.6 Martian days in a Martian year, and the sun moves through one constellation of the zodiac in about 56 Martian days.

These solar phenomena are slower than on earth, but the lunar phenomena are much faster and full of dynamic energy and variety. There are two small moons, not large enough to cover the sun in an eclipse. The largest, Phobos, moves eastwards completely round the zodiac more than three times in the course of a sidereal day, and therefore rises in the *west* and sets in the east. Casting a dim light, it appears about twice as large when high in the sky than when near the horizon, as it is so close to the surface of Mars. For an observer near the Martian equator, Phobos spends little more than four Martian hours above the horizon but stays longer beneath it, rising about every eleven hours. In a day (24 hours 37 minutes) it completely crosses over the Martian sky twice while passing through its phases three times. Each time it appears above the horizon it is in a different phase cycle to before, each cycle taking less than one-third of a day. It has a 'month' of 7 hours 39 minutes. It can appear full twice in a night from a fixed location on Mars' surface (Figure 9.25). Phobos is so close to Mars that this lively display cannot be seen from the polar regions of the planet beyond 70 degrees north or south.

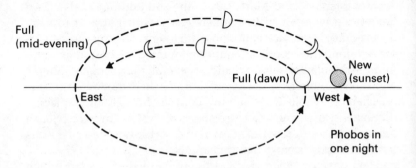

Figure 9.25 Phobos in one night

Solar 'eclipses' by Phobos are very frequent, nearly 1,400 occurring in a Martian year. The moon would appear as a black area transiting the sun's face, only covering the solar disc to a small extent. Eclipses of Phobos by Mars' shadow would number over 1,300 a year. Seen from Phobos, Mars with one edge on the horizon would have a diameter extending half-way from the horizon to the zenith. Mars' relation to the horizon would remain fixed and the planet would appear to rotate in about eleven hours and go through its phases in just over seven hours.

The smaller and more distant moon Deimos has quite opposite behaviour. It rises in the *east* and sets in the west very slowly, taking more than two days (fifty-eight Martian hours) to accomplish this journey and can become full, seen from the Martian equator, twice while thus above the horizon (Figure 9.26). Between two risings Deimos passes through more than

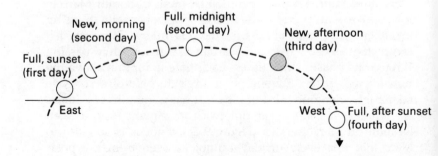

Figure 9.26 Deimos between rising and setting

four 'months' of phases, but most of this time is spent beneath the horizon.

Being both smaller and more distant than its companion moon, Deimos appears in the sky like a bright planet, fluctuating in brightness with its phases. It has about 130 each of 'lunar eclipses' and solar transits per year and there are rare occasions when Deimos is totally eclipsed by Phobos. On other occasions both moons can pass together across the face of the sun or both be eclipsed simultaneously by Mars' shadow. Seen from Deimos, Mars would remain stationary above the horizon and have a diameter of about half a zodiac constellation. Phobos would appear in all phases, passing across the face of Mars or being occulted by it in a variety of phases.

These two moons were discovered in 1877 by the American astronomer Asaph Hall, and named Phobos (Terror) and Deimos (Panic) – attendants of Mars, after Homer's *Iliad* (Book 15) which describes the god Mars setting out to avenge the death of his son Ascalaphus: 'And he called to Terror and Panic to harness his horses, while he himself put on his glittering arms.'

In 1610, after Galileo's discovery of four moons of Jupiter, Kepler stated on a mathematical basis that Mars would be likely to have two moons. Apparently taking up Kepler's idea, Jonathan Swift gave an astonishingly good approximate description of the moons in *Gulliver's Travels* 150 years before they were discovered. Swift wrote that the advanced Laputan astronomers discovered two satellites of Mars 'whereof the innermost is distant from the centre of the primary planet exactly three of his diameters, and the outermost five; the former revolves in the space of ten hours, and the latter in twenty-one and a half'.

The world of Mars has long fired the imagination of Man. It still can if we place ourselves in thought upon its surface and sense the very different astronomy to ours which an inhabitant would experience. The picture would include our earth taking to the skies as an interior planet and shining as a bright morning or evening star. Periodically the earth would transit across the face of the sun, this event taking place (on most recent dates and dates close in the future by the earth calendar) on 8–9 November 1800; 12–13 November 1879; 8–9 May 1905; 11 May 1984; 10 November 2084; and 14–15 November 2163. They occur in either April–May when Mars is near the descending node of its orbit, or October–November when near its ascending node. Mars, of course, is in opposition as seen from the earth when these transits take place. The transit of 1984 (Figure 9.27) had a duration of 8½ hours.

Figure 9.27

Finally, we can bring all the naked-eye planets together as seen from earth and note the phenomenon of planetary gatherings. Between the years AD 1007 and AD 2100 there are fourteen occasions on which the sun, moon and five planets occupy an area of the zodiac less than 30 degrees wide as seen from earth. Strangely, there will be no other gatherings of this sort for more than 300 years after 2100. The most compact grouping took place in September 1186 among the stars of the Virgin within an angle of 12 degrees. The next most compact grouping was on 5 February 1962, when the planets came within a 16-degree angle in the constellation of the Goat. This occurrence was exceptional in that the sun was at the same time totally eclipsed by the moon, so that the planets could be seen in a dark sky ˙ from any place within the moon's umbra. The shadow track started in Borneo and moved over New Guinea and across the Pacific, ending short of the west coast of the USA. Observers in New Guinea saw Venus, Mars, Jupiter and Saturn during the eclipse. Mercury was just above the sun and too close to it to be observed. Normally these gatherings are not observable at all because they include the sun.

The next grouping will take place on 5 May 2000, when the planets will lie within a 26-degree span in the constellation of the Ram (Figure 9.28). Two more follow, both of 29 degrees and both in the constellation of the Virgin, in September 2040 and November 2100.

In this chapter the principal aim has been, firstly, to open the way to a qualitative knowledge of the naked-eye planets

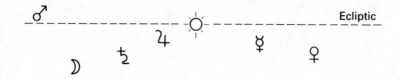

Figure 9.28

through the phenomena, and secondly, to distinguish the essential difference between interior and exterior planets as experienced from earth. For instance, after a while one perceives a certain quality of youth in the interior planets and one of older age in the exterior. This thought also occurred to Ptolemy who, in his book the *Tetrabibilos* described the ages of Man in terms of the planetary spheres from moon to Saturn. The first four years of life are adapted to the moon (the moon having four quarters, he explains); age four to fourteen is adapted to Mercury; fourteen to twenty-two corresponds to Venus (its cycle of synodic periods being eight years); twenty-two to forty-one is subject to the sun (he called nineteen the sun's number, perhaps in connection with the Metonic Cycle); forty-one to fifty-six was claimed by Mars (synodic cycle of fifteen years); fifty-six to sixty-eight came under Jupiter (synodic cycle of twelve years); and the age of sixty-eight onwards was regulated by Saturn.

Such is an ancient reading of the qualitative script of the stars. Commenting on the *Tetrabibilos*, the science historian George Sarton said, in his *Ancient Science and Modern Civilisation* –

> One cannot read the whole of that treatise or a part of it
> without being terribly dismayed. If Ptolemy was really
> the author of it, it is a thousand pities, but that only
> shows that he was a man of his clime and time. Even
> the greatest genius cannot transcend all those limitations
> at once.

If the greatest genius cannot transcend all the 'limitations' of an imaginative age, then neither can Dr Sarton transcend all the limitations of a cynical one.

Chapter 10
Comets, Meteors and New Stars

Comets, meteors and new stars have in common the fact that they are temporary phenomena and are often, if not always, unexpected apparitions in our skies, crossing the threshold of the visible and the invisible worlds.

Dramatic appearances of these phenomena easily take on the character of announcements to the imagination. Comets have a long history of being associated with human affairs, accompanying positive or negative events. They were considered to be coincident with disasters, or the births or deaths of important people. Giotto painted a comet above the head of the infant Jesus in a fresco in the interior of the Avena Chapel in Padua. Shakespeare's *Julius Caesar* contains the well-known lines:

> *When beggars die, there are no comets seen:*
> *The Heavens themselves blaze forth the death of princes.*
>
> *(Act II Scene ii)*

In the same year that King Harold of England was defeated and killed at Hastings in 1066, an early appearance of Halley's Comet stood in the sky and was woven into the French Bayeux Tapestry which commemorated the invasion of William the Conqueror. One picture shows a group of Englishmen pointing at the comet, and the text reads: 'They are in awe of the star.' John of Damascus, a Greek Church Father, wrote 'It often happens that comets arise. These . . . are not any of the stars that were made in the beginning, but are formed at the same time by divine command and again dissolved.'

Giotto's comet is remarkable for its naturalism, and evidence points to his using as a model the 1301 appearance of Halley's Comet which he almost certainly would have seen. It is understood that Giotto's Avena frescoes were painted in 1303/4.

Giotto gives the nativity comet a red colour and places at its centre point, or nucleus, an eight-pointed star. The nucleus of a comet is not actually visible, but there is often a bright area near the centre of the surrounding coma. Together they make up the 'head'. Thus many comets have a threefold appearance of coma, centre point and tail (Figure 10.1). The tail frequently

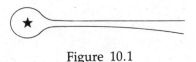

Figure 10.1

has two aspects – a straight edge or line, and a curved part. Light from the straight part is luminous and bluish, from the curved part is reflected sunlight as in the case of the nucleus, and the coma has both properties. Some comets have only straight or curved tails while others combine the two. Many have no tails at all. Comet IRAS-Araki-Alcock in May 1983 glowed as a ghostly, circular patch in the northern heavens, turning through the night round the Pole Star with the Great and Little Bears. Even such a modest, tailless appearance impressed one with its uncanny presence intruding upon familiar skies – a nebulous amoeba drifting by.

It is, in fact, widely understood that comets comprise attenuated, elementary substance arising from the formation of the sun and planets, and therefore represent solar system matter at an early stage of evolution – visitors from the past. A European Space Agency mission called 'Giotto' (commemorating the artist's 'nativity comet' painted in Padua) will send a spacecraft to intercept Halley's Comet in 1985/6 and enter its coma in an attempt to learn more about the physical nature of these strange visitors from space. Two spacecraft from the USSR and two from Japan will also approach the comet.

The diameter of Comet Halley's coma will be many times that of the earth. Some comet heads grow to be larger than the sun. Yet one cannot speak of a comet head or tail as being physical in any ordinary way. The density would be something equivalent to a dozen marbles scattered through an area of a cubic kilometre. In 1910 the tail of Halley's Comet may have swept across the earth but nothing was detected.* Stars shine undiminished when seen through the tail of a comet as if it were not there. Even astronomers this century have classified the nucleus as no more than an 'apparent phenomenon'.

Comets are insubstantial veils of luminescence and reflected sunlight with hardly any physical body at all. Yet they cause the greatest dramatic appearance of all celestial phenomena. Their characteristics incline them to appear from any point on the celestial sphere; to grow into visible existence as they approach the sun; to respond to the sun with, very often, extending tails, huge comas and brightening nuclei; to curve round the sun before shrinking in size, withdrawing their tails, extinguishing their luminescence and disappearing when they have gone beyond about twice the sun's distance from the earth. The tails of comets are always turned away from the sun, as if something is being syphoned off, and can stretch as far as the distance between earth and Mars, and longer.

Figure 10.2 shows the simplified path of a comet. Comets travel, in detail, in very complicated, three-dimensional curves due to deflection (perturbation) by large planets such as Jupiter. If there were no deflections and comets were attracted to the sun only, they would move in planes in, mostly, parabolas. A parabola is the curve shown in Figure 10.2 and its two arms

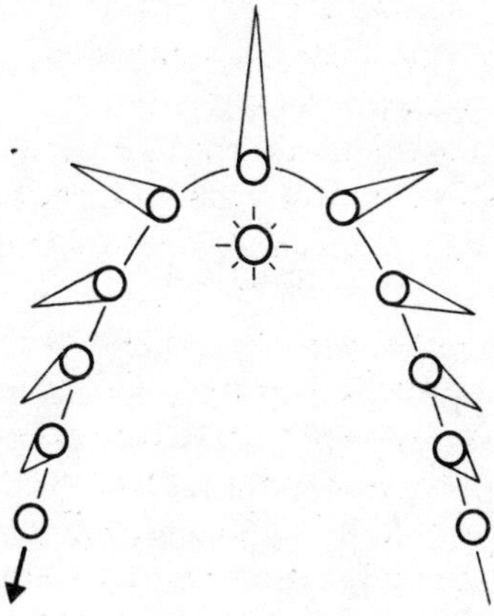

Figure 10.2

widen into space and only come together (mathematically) at
infinity. So the comet would not return. Most comets are
deflected into ellipses (Figure 10.3) and do return. They can
also be turned into hyperbolas which widen out further than
the parabolas (Figure 10.4) and cannot return either. Kepler

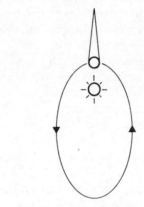

Figure 10.3

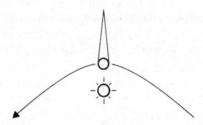

Figure 10.4

thought that comets moved in straight lines, which is not incom-
patible with their appearance on the celestial sphere and which,
in fact, retains the essence of the parabolic, non-returning side
of their nature. Kepler pioneered the thought that a straight line
has only one point at infinity, which is now part of synthetic, or
projective, geometry as taught today. Both straight line and
parabola, in one sense, emerge from and return to their same
points at infinity.

But most comets are deflected into elongated ellipses which
extend beyond the planets and their average time of return has
been estimated at some 40,000 years. At each return a fresh

coma is formed as it nears the sun. The shortest-known return period is 3.3 years and belongs to Comet Encke, first seen in 1786. Periodic (returning) comets take on a planetary character and approach becoming normal members of the planetary system, their calculated orbits displaying loops all the way round the celestial sphere when they are projected against the stars. Figure 10.5 shows the one-loop-per-year progress of

Figure 10.5

Halley's Comet from its first detection on 16 October 1982, near the star Procyon in Canis Minor. The comet was then a little further out than the planet Saturn. Its direction and speed were close to predictions, which meant that the comet would arrive only half a day earlier than expected at its closest approach to the sun (perihelion) on 9 February 1986. Many comets with calculated periodic orbits, however, are much more difficult to predict on their return and sometimes do not return at all.

As seen, Comet Halley moves in a retrograde direction (westwards) against the normal movement of the planets. This is unusual for a 'short-period' comet, i.e., one which returns within about 200 years. Almost all short-period comets have direct motion, as do the planets. However, of the long-period type, as many have retrograde orbits as have direct ones.

Since a probable Chinese sighting in 240 BC, Halley's Comet seems to have returned at a mean period of almost seventy-seven years. In 1985/6 its appearance will not be as favourable as in 1910 – at perihelion it will be on the far side of the sun as seen from earth, below the equator and among the stars of the Waterman. In the northern hemisphere it will be best seen at the beginning of 1986 in the evening sky, and best seen in the southern hemisphere after perihelion in the morning sky. However, it will also appear in evening and morning skies in both hemispheres and Figure 10.6 shows positions seen from geographical latitude 50 degrees north about half-an-hour after sunset when the brightest stars are visible. Figure 10.7 shows

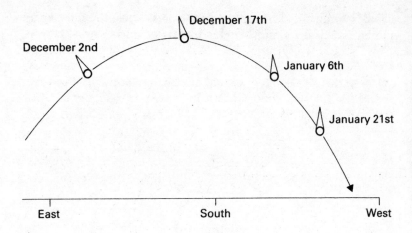

Figure 10.6 Evening sky

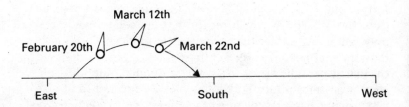

Figure 10.7 Morning sky

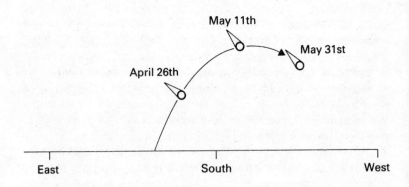

Figure 10.8 Evening sky

the morning appearance from the same latitude. At the end of April it will enter the northern hemisphere evening sky once again as it recedes, moving upwards above the southern horizon (Figure 10.8).

During mid-January, the comet will stand among the stars of the Waterman in the evening. It is interesting to note that exactly twelve years earlier comet Kohoutek stood in almost the same position in the evening sky, and as Jupiter has a twelve-year cycle, this planet will be placed near Comet Halley in 1986 as it was near Kohoutek in January 1974. Venus was an evening star in this area of the sky in 1974, but will be absent in the middle of January 1986, replaced by a crescent moon. Venus will be near superior conjunction with the sun, but will rise to join Jupiter in the evening sky after Comet Halley has moved the other way to reach sun conjunction on 9 February.

Like Comet Halley, Kohoutek had perihelion on the far side of the sun (28 December 1973) but moved in the normal 'direct' motion against the stars, following planetary motion. Another difference is that Kohoutek becomes inherently brightest before perihelion, while Halley tends to do the reverse. Therefore both approach maximum brightness on the same side of the sun (west). Also, this is understood to be Kohoutek's first appearance and it is not expected to return for something like a million years, if it returns at all. So the picture is of a new, direct-motion comet opposing the path, in the stars of the Waterman, of an old, retrograde-motion comet, with Jupiter in attendance. Kohoutek anticipated Halley and even prepared the public for Comet Halley's return. Although it turned out to be disappointing visually, Comet Kohoutek created considerable worldwide interest in 1974 and stimulated the imagination concerning comets and astronomy in general. This was a precursor to the same thing happening twelve years later, with attention focused on the same constellation of the zodiac, on the same evening skies in January.

It is estimated that there are about twenty to thirty naked-eye comets every century, but these are only a fraction of the number which continually draw in towards the sun. Kepler was on the right lines when he suggested that there were probably as many comets in the heavens as fish in the sea. When visible, they can remain so from a few days to more than a year. During this time most comets seen in recent history were white in colour, though there are many records of coloured comets, particularly in early accounts. They have been reported as being blue, bluish, red or reddish-yellow, gold, and greenish. It is difficult to know how much these descriptions represent the comet's own colour or the effects of the atmosphere. Comets can also appear in the most varied shapes. Multiple tails are possible; for example, de Chéseaux's Comet of 1744 had six tails spreading out like a fan, and more have been noted on other occasions. A sixteenth-century manuscript, attributed to Nostradamus, pictures and describes the effects of nine comet types of varying colour and shape, from star-like with the complexion of the moon, to red and sword-shaped. Probably the brightest of modern times was the Great Comet of 1843 which was discovered during daylight. It passed very close to the sun and developed a tail over 70 degrees in apparent length, extending in space the distance between the sun and Mars.

The word 'comet' comes from the Greek 'aster kometes', meaning a long-haired star. This suggests a feminine nature. According to Greek writers, the Egyptians before them made the analogy between female tresses and the appearance of comets. The Chinese called them 'broom stars' and a Chinese description of a comet of 524 BC saw it as a 'new broom' to sweep away traditions and the old order of things.

The origin of comets is still a matter for conjecture and theorising, and they remain something of a mystery. It is usually considered today that they arise from within the planetary system and do not enter it from the realm of stars. Yet these wandering bodies are so delicate that half-way to the nearest star is still an area in which, technically, they can stand within the planetary system and be induced to approach the sun. Many theories in the last two hundred years have been put forward to explain their origin and these fall into two main groups – the concept that there is a huge cloud of 'captured' inter-stellar matter at far distances in space under the influence of the sun; and the concept that cometary material is first ejected by bodies of the solar system.

One development of these theories is that the distant 'cloud' originated from the disintegration of a planet which once orbited between Mars and Jupiter. This orbital area now contains thousands of asteroids or minor planets which are too small to be seen by the naked eye. Other theories have suggested that comets are the result of ejections from planets like Jupiter and Saturn.

A once widely accepted theory was that cometary material is thrown off from the sun, but doubts arose concerning possible trajectories for this to result in the paths of comets as observed. Yet there is a case for no possibility being ignored when dealing with phenomena whose physical properties are not yet fully understood. Science has been surprised too often. As late as the 1950s the Cambridge astronomer R. A. Lyttleton proposed that comets resulted from a process of accretion, near the sun, of inter-stellar matter. The sun thus created the comets and impressed on them their orbital forms.

On an imaginative level, the unpredicted appearance of a new comet is like a Whitsun event among the stars – a sudden flame from the depths of space, arriving from, and departing to, the unknown. The very nature of comets is such that we continually seek to know whence they come and they remain an enigma in our skies, prompting questions on the origin and formation of matter. Yet the comet itself has, in modern times, been pointed to as the origin of another elusive celestial phenomenon – the meteor flash.

In 1861 the American astronomer Kirkwood suggested that meteors were brought about by contact between the earth's atmosphere and fine debris scattered along the orbits of comets. This theory is still held in favour today. The first observational connection came in 1866 when the Italian astronomer Schiaparelli announced that the famous Perseid meteor shower, which flashes forth every August in the constellation of Perseus, was associated with the orbit of Comet Swift-Tuttle (1862). This annual display, the most abundant of the regular showers, is traditionally known as 'the tears of St Lawrence' as it takes place around 10 August when the Christian martyr is commemorated. The occurrence of these meteors can be traced back to the tenth century and they have long been recorded in the church calendar of England. The orbit of Halley's Comet is associated with the Orionids of October and the Eta Aquarids of May.

It so happens, though not for any technical astronomical reason, that the distribution of the major showers in the year results in visible meteors being more abundant between August and January. This has no connection with falls of, for example, iron meteorites on the earth. The common meteor is a phenomenon of light which is not observed to produce a 'meteorite' fall. It is estimated that the particles which cause meteor flashes are as small as grains of sand or, at most, pebbles when they encounter the earth's atmosphere and burn up. Meteorites discovered on the earth, on the other hand, have been observed to originate from fireballs – great flashes brighter than Venus which are stray occurrences, not associated with normal meteor showers. The origin of meteorites is often considered to be the asteroid belt, though comet material is not ruled out in some cases and even the planets and the moon have been suspected as sources. But the common 'shooting star' and meteorite falls are not seen to be connected with each other.

A shower of meteors can simply mean the appearance of one meteor flash per minute or, more often, less than that. The streaks across the sky from one family or shower seem to radiate from one source on the celestial sphere which can be plotted to a small circular area. This is called the radiant, and a rough tracing-back along the paths of flashes in a shower indicates a radiant point. This is, in fact, the effect of perspective. Just as sunbeams descending from a cloud are virtually parallel though they appear to converge, so the separate streaks in a meteor stream are also virtually parallel (Figure 10.9). The particle

Figure 10.9

striking the earth's atmosphere and causing the streak is relatively close – averaging about 150 kilometres – but is projected from the point of view of the observer onto the celestial sphere. This results in the direction of the streak being different for different observers but the radiant area being more or less the same. Figure 10.10 shows a meteor path (m) which becomes a

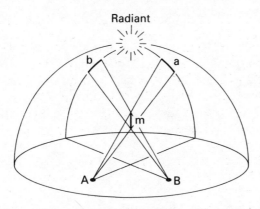

Figure 10.10

flash against the stars (a) for an observer at A, and the flash (b) for an observer at B. Strictly speaking the two observers would experience slightly different celestial spheres from their separate positions, but the difference would be negligible for the purpose of this illustration. An observer standing immediately within the flight path would see a point light up in the sky as the meteoroid travelled directly towards him.

Most meteor flashes are reported as being 'straight', but some are said to be curved. Certainly a long flash appears, to the author, often to be curved. Some bright ones have been reported to change direction and even move on the arc of a circle. Usually the colour of the luminous streak is described as white, though there are descriptions of greenish, reddish or yellow trails. Colour has been noticed to vary with the height and calculated velocity. A fireball (producing a fall of meteorites) at Pultusk in Poland in January 1868 was observed to begin its appearance as an ordinary meteor, then its light increased and the colour changed to a bluish green, then towards red. In the case of a fireball the increase in intensity of light at one stage has the effect of making the trail seem to broaden out then narrow again, lending the appearance of a spear (Figure 10.11). A fireball is sometimes called a 'bolide' from the Greek word meaning a thrown spear. Some of their trails remain hanging in the sky for long periods, even for hours. The advent of a fireball is often said to be accompanied by sounds described as singing, buzzing, crackling, hissing or rustling.

In the town of Helston in Cornwall there is a stone, built into

Figure 10.11

the wall of the Angel Hotel, which local legend says was cast into the centre of the town centuries ago by a dragon. The patron of Cornwall and Helston is Archangel Michael, the dragon slayer, and between the eleventh and eighteenth centuries meteoric fireballs were referred to as flying dragons. An old engraving of another fireball in AD 1000 shows, along-side the fiery descent trail, a dragon with a luminous head and blue feet careering headlong down the sky.

The German scientist Alexander von Humboldt (1769–1859) first suggested that meteors in a shower might originate from the same position in the sky (radiant). In 1799 Humboldt and his companion Bonpland, while on a journey to South America, rose early one morning to enjoy the air and, quite by surprise, witnessed a magnificent display of hundreds of thousands of meteors lasting over two hours. These were later identified as the Leonids, emanating from an apparent position in the constellation of Leo. The suffix 'id' attached to a constellation name to designate a meteor stream comes from the Greek, here meaning 'children of Leo'.

The greatest-known Leonid display was on the morning of 12 November 1833, observed from the West Indies to Canada. At the peak of the shower one witness was reminded of a snowstorm and 'one thousand meteor flashes might be counted every minute'. The Leonids also provided what is understood to be the greatest meteor display ever recorded. This was on the morning of 17 November 1966, and also visible from America. More than 2,000 meteors per minute at the peak of activity were estimated. These spectacular displays of Leonids tended to recur in periods of about thirty-three years and the shower has been identified with a comet of 1866, discovered by Tempel and Tuttle, which has the same period (due to return in 1998). Another Leonid meteor 'storm' is anticipated for 18 November 1999, though the occurrence is not certain. Only a few were seen in 1899 and 1933. In between the predicted special displays, the Leonids, like every other shower, recur every year, though modest in number. In fact, Leonid activity is normally feeble.

The Leonids are the swiftest of meteor phenomena in their path across the sky. The morning hours bring the swiftest meteors of any shower and are the time when visible meteors are most abundant. The technical maximum is 6 a.m. (often earlier visually due to dawn light) when about three times as many occur than in the evening. Also, the meteors after sunset tend to be slowest. Disassociated from the showers (or seemingly so) are the sporadic meteors – the outsiders which appear from any direction at any time. An average of about seven sporadic meteors an hour can be seen on any one night.

An additional aspect of frequency is that in the course of the year, most meteors are seen on autumn mornings when the ecliptic stands high against the horizon. Also, as it happens, most radiants are in the northern half of the celestial sphere, favouring observation from the earth's northern hemisphere. However, meteor radiants do not remain exactly stationary in the sky. From night to night they shift about 1 degree eastwards on the celestial sphere due to orbital elements. If not affected by other disturbances, the date of a shower's peak will move later in the year by about one day in seventy-two years due to precession of the equinoxes. A shower will also vary in the time of day it reaches maximum due to the calendar year not being a whole number of days.

If a shower takes place during daytime, it can nowadays be detected by radar. This also detects meteors invisible to the naked eye day or night. Daytime showers are most common in June and it is estimated that thousands of millions of meteors encounter the earth's atmosphere every day, bringing with them thousands of tons of material. Then there are the micro-meteorites which cause no flash but slowly permeate the earth's environment from floating as dust in the outer reaches of the atmosphere to sinking to the ocean floors.

Meteor phenomena form a direct connection between the

earth and the region beyond, flashing into view at the physical periphery, plunging to earth or sea in the form of solid meteorites, or sifting through atmosphere and water in homeopathic quantities. The upper regions thus penetrate the lower. When one gazes on the face of the night sky, surprise and awe arise at the sight of a meteor slipping silently across the stars. One involuntarily becomes immobile and turns all attention upwards, waiting for the next flash and extending the senses into the darkness which has acquired new mystery. The meteor is an intuition, flashing across the cranium of the sky.

Meteors are those innovations which move swiftly among the stars, comets are those which move slowly, and new stars are those which move not at all. New stars appear steadily in the realm of the other fixed stars – announcements of events on the grandest scale.

Modern astronomy distinguishes two types of new star – novae and supernovae. To the naked eye both of these may well appear the same, but radio astronomy after the Second World War identified the supernova as a once-only event marking the cataclysmic end of the life of a star with its explosive brightening, and the nova as an outburst on a smaller scale which may, theoretically, be repeated. Novae are the more frequent phenomena, six bright naked-eye occurrences having taken place this century in 1901, 1918, 1925, 1934, 1942 and 1975. The term supernova was coined in 1937 and the last known phenomenon of this kind in the visible star system (galaxy) was in 1604.

New stars feature in Chinese records dating back to before Christ. Few records of such events occur in Europe and the Arab countries. A list of seventy-five new stars derived from reasonably reliable sources and published in 1976 by Richard Stephenson of the University of Newcastle upon Tyne, contains only three such events observed in the West – those of AD 1006, 1572 and 1604. The first sighting listed by Stephenson is a Chinese one in the spring of 532 BC, followed by seven others from the same nation before the time of Christ.

The Chinese called them 'guest stars' and they figured strongly in astrological predictions concerning affairs and personages of state. The following description relates to a new star which shone for five months in the year AD 369, high in the northern sky: 'A guest star was seen at the western wall of [the constellation of] Tzŭ-wei . . . The interpretation when a guest star guards Tzŭ-wei is assassination of the Emperor by his subjects. In the 6th month Huan-wei dethroned the Emperor.' For the Chinese the earthly and the heavenly king-doms reflected each other, and star groups were given such titles as: Emperor; Crown Prince; Secretaries; Court Eunuchs; Celestial Bed; Guest Houses; Inner Kitchen, etc.

The three new stars mentioned already which have been observed in the West in the last thousand years appeared in or beside the Milky Way – the richest field of stars, with an area in the constellation of the Archer which is the most dense. Today this densest area is conceived of as the centre of our galaxy. All stars visible to the naked eye are part of this galaxy – a lens-shaped multitude of stars when pictured edge-on, and made up of a spiral form when pictured in plan view. Our sun and earth, if placed in the middle plane of this lens and some distance from the centre, have a view which shows the main part of the lens as a milky band of distant star glow encircling the sky. Like the zodiac, this band is tilted to the earth's celestial equator and goes through movements in relation to the horizon throughout the seasons in a similar way to the zodiac. Figure 10.12 indicates some of the constellations lying within or near

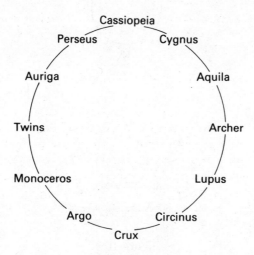

Figure 10.12

the Milky Way. This includes stars in the southern hemisphere as the Milky Way passes not far from the north and south poles, crossing the zodiac twice (translated names are retained for the zodiacal constellations, as elsewhere in the text).

On autumn evenings in mid-northern latitudes the Milky Way stretches from due east up to overhead and down to due west, with Cassiopeia in the overhead position. On spring evenings the Milky Way lies along the northern horizon from east to west. On summer and winter evenings it stands high in the sky again, this time meeting the horizon in north and south. This cycle of positions is moved through, of course, in the space of twenty-four hours, but unlike the zodiac we only see part of the complete Milky Way circle, one part being among the southern hemisphere stars beneath the horizon. The section we see has one point, for example, at the star Deneb in Cygnus which makes the movement shown in Figure 10.13 at latitude 52 degrees north. Figure 10.14 gives a diagrammatic picture of the four positions mentioned above. It is a misty, insubstantial type of zodiac which weaves all round the horizon (in azimuth) and from highest to lowest (in altitude).

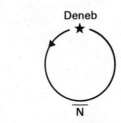

Figure 10.13

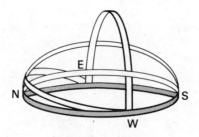

Figure 10.14

The word 'galaxy' comes from the Greek for 'milk'. The Milky Way was their 'circle of the galaxy'. It was milk dropped from Juno's breasts while she suckled Hercules. The Greeks also called it 'the road to the Palace of Heaven'. Many cultures have connected the Milky Way with the concept of a river or road. The ancient Akkadians called it 'the River of the Shepherd's Hut' and 'the River of the Divine Lady'. Both Greeks and Romans saw it as a pathway for departed souls who entered by the door where the Milky Way intersects the zodiac in the Twins and left it to return to the gods by the door in the Archer.

Such is the background within which were placed the two most renowned new stars (supernovae) of recent history which blazed forth from the Milky Way within the phenomenally short time of thirty-two years of each other. The first appeared in 1572 in the high-climbing constellation of Cassiopeia, and the second in the low-lying southern tip of Ophiuchus which borders on the zodiac near the 'door' of the Archer. Each arrived fortuitously during the lifetimes of two of the greatest astronomers whose observations of these phenomena opened the way to modern stellar astronomy. The earlier is known as the new star of Tycho Brahe and the later as that of Johannes Kepler. What was established was that the new stars shone from the distant region of other fixed stars, overturning the old Greek view that this was a changeless region.

In November 1572 Tycho Brahe, a twenty-five-year-old Danish nobleman, was living with his uncle at the Lutheran Abbey of Heridsvad, about 20 miles east of Helsingborg. Brahe had set up a laboratory in an outhouse of the abbey and was applying himself to chemical experiments, chemistry or alchemy being considered as a science which was an integral part of the cosmic order of things. In the study of the metals, silver was related to the moon, quicksilver to Mercury, copper to Venus, gold to the sun, iron to Mars, tin to Jupiter and lead to Saturn.

On the evening of 11 November 1572, Brahe was returning from the laboratory to the house for supper, when he noticed a bright star high overhead where, from his knowledge of the constellations, no star should be. He was so astonished that, hardly believing his eyes, he turned to some servants who were accompanying him and asked whether they saw it. Even though they answered that they did, he called out to some country folk who were driving by and asked the same question. They gave the same answer and by this time he could believe his own senses. The star had added itself to those in the constellation of Cassiopeia as shown in Figure 10.15 and was brighter than any of them.

★ New star

★

★

★

★

★

Figure 10.15

Brahe had just finished making a new measuring instrument for the stars, a sextant, which afforded greater accuracy than his previous one (a cross-staff). That night he measured the distance of the new star from the others in Cassiopeia, then waited for the next night to see if the star was still there and, if so, whether it had changed its position. Its position did not change and, in fact, it remained visible for about eighteen months.

This constellation is circumpolar, so he could, over an interval of time, plot the new star's position on a circle round the pole. If the new star was nearer to the earth than the others (that is, beneath the changeless celestial sphere) then this would be revealed by 'parallax'. In other words, the position of the new star would appear to shift against the distant stellar background as it moved round the pole. This was a most important aspect, the answer affecting the whole future of astronomy and its connection with theology.

The principle of daily parallax can be simplified, from Brahe's point of view, as follows. In Figure 10.16 let E be the centre of the earth, S a fixed star, N a new star and O the observer. If the new star is closer to the earth than the fixed star, then the angle NOS must be larger when the stars are near the horizon than when high above it and therefore when near the horizon the stars will appear further apart. This expresses what is called horizontal parallax.

Brahe carefully measured the angle between the new star and the star Schedar in Cassiopeia when both were at highest and lowest positions (upper and lower culmination on the

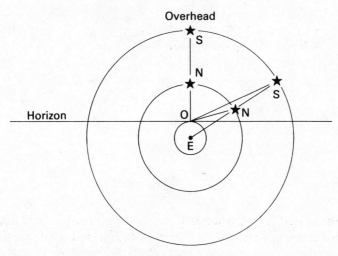

Figure 10.16

meridian). He found no difference in their distances apart in either position. Absence of parallax for a new star, shown clearly for the first time in history, established that the 'eighth sphere' or realm of fixed stars was subject to change. This discovery was at least equal in importance to the theory of Copernicus regarding Man's concept of the universe and opened the way to modern astrophysics. This is despite the fact that the new star of 1572 pre-dated the telescope and all the sophisticated instrumentation which later cast a fine trawl-net over the sky to trap celestial objects into the confines of a new philosophy.

Tycho Brahe, however, was a thinker who retained a belief in a stationary earth. In a brilliant stroke he devised a cosmic system in which the earth stood at the centre, moon, sun and stars moved round the earth, and the planets moved round the sun – thus uniting the Copernican theory and the geocentric experience (Figure 10.17). Calculations and observations for the Tychonic system were identical to that for the Copernican. He made the orbit of Mars cross that of the sun as he believed, mistakenly, that at opposition Mars came closer to the earth than the sun's orbit.

At first, Brahe did not want to publish anything on the new star, partly because it was not proper for a nobleman to write books. But he was persuaded to do so by friends and by wildly

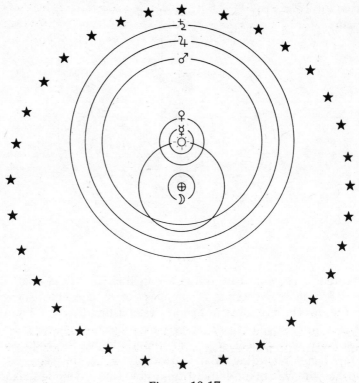

Figure 10.17

inaccurate German accounts of the star's position which placed it at a distance of only twelve or fifteen semi-diameters of the earth. His description was combined with an astrological and meteorological diary which he had prepared for 1573, and was given the title *De Nova Stella*. In this book he said that nothing similar to the new star had been seen since Hipparchus reported one in about 125 BC and which many had taken to be a comet. No similar star had been seen since, for the star of the Magi was not a celestial object, he said, but something relating exclusively to them, and only seen and understood by them.

The new star, he continued, twinkled like other stars, while the planets did not twinkle, which was another proof of its belonging to the eighth sphere (see Figure 12.18). Referred to division lines of longitude through the poles, the new star belonged to the sign of Aries (the astrological sign measured

from the spring equinox, not the star constellation of the Ram –
Figure 10.18). He pointed out that historically the star appeared

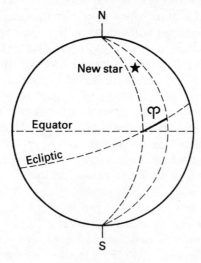

Figure 10.18

almost at the conclusion of a 'watery trigon' in the sign of Pisces
and the commencement of a 'fiery trigon'. This is connected
with conjunctions of Jupiter and Saturn discussed in Chapter
9. A conjunction at the end of a cycle in watery signs was to
take place in 1583, and one at the start of a fiery cycle in
Sagittarius in 1603 (Figure 10.19).

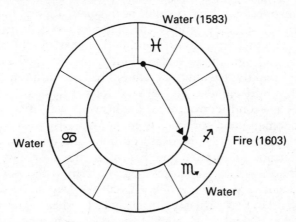

Figure 10.19

In a later book, called *Introduction to the New Astronomy*, Brahe said that the movement of the trigon from watery to fiery signs (in which Aries, the first sign of the zodiac, stood) began an 800-year period within which would be inaugurated a peaceful period when the lion will eat straw like the ox as foretold by the prophet Isaiah (chap. II, v. 7). He said that the place on earth from which this change would arise was that place which had the star at its zenith (directly overhead) when it first appeared to announce the new age. He assumed this to be at the time of new moon previous to when he first saw the star, and calculated the spot to be 'in Russia or Moschovia where it joins the north-east part of Finland'.

Yet the star itself, apart from the Jupiter-Saturn cycle, had its own particular, though shorter, influence with which to introduce the new historical period. Brahe wrote that the star of Hipparchus announced the extinction of Greek ascendancy and the rise of the Roman empire. The star of 1572, he said, was the forerunner of vast changes, not only in politics but also in religious affairs. It had shone forth from the spring quarter of the sky and therefore announced that some great light was at hand, and as it was visible over most of the earth, so the effects of it would be felt over the greater part of the globe.

Brahe considered that the new star became visible because it was illuminated by the sun and was formed from the substance of the Milky Way. In his earlier book, *De Stella Nova*, Brahe had said that at first the new star had shone like Venus and Jupiter and its effects would therefore be pleasant; but as it then became red like Mars, there would next come a period of wars, seditions, captivity, and death of princes and destruction of cities together with dryness and fiery meteors in the air, pestilence, and venomous snakes. Lastly, the star had shone like Saturn and there would, therefore, finally come a time of want, death, imprisonment, and all kinds of sad things.

Yet a second innovation in the heavens was in store for the astronomers and prognosticators. With astonishing coincidence, shortly after the death of Brahe and one year after the conjunction of Jupiter and Saturn in the new fiery trigon, another new star (supernova) blazed forth in the sky. Brahe's star stood 'high' in the Milky Way, half-way along its visible arch and in that part of it which climbs overhead. The next new star stood in the lower part of the Milky Way among the stars of Ophiuchus, near the zodiacal constellation of the Archer. But the most astonishing part of the coincidence was that this second new star appeared in the same region of the sky as Jupiter and Saturn which had recently been in conjunction, *and* into which Mars had also moved.

The three exterior planets and a new star stood together. Brahe was dead but his successor and former co-worker Kepler was just approaching the age of thirty-three. The star of 1572 had appeared a year after his birth. Kepler was now imperial mathematician to Emperor Rudolph II in Prague. On 9 October 1604, Mars was in conjunction with Jupiter and on the following day an amateur astronomer among the court officials looked through a gap in the clouds to observe the planets. To his surprise there was a bright new star shining just above them. At dawn the next day he went to Kepler with the news, which Kepler hesitated to believe. The evening skies remained overcast until 17 October when Kepler saw the spectacle for himself (as shown approximately in Figure 10.20). The new star

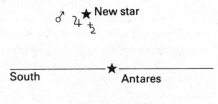

Figure 10.20

competed with Jupiter in brilliance. After publishing a short report right away in German, Kepler produced a Latin treatise *De Stella Nova in Pede Serpentanii* two years later, after the star had moved into the morning sky, back into the evening sky, and dimmed to disappearance in October 1605. In Italy, reports of the star described it as 'like Mars' and 'like half of a ripe orange' while a Chinese account stated: 'At the beginning of the night in the south-west there arose a strange star as large as a crossbow pellet. Its body was orange in colour. It was called a Guest Star.' A Chinese account of Brahe's star of 1572 also described it as 'like a crossbow pellet', adding that the

Emperor of Shên-tsung 'saw it in his palace. He was alarmed and afraid, and at night he prayed in the open air on the Vermillion Steps'.

For the 1604 appearance, Kepler, in the Latin text, begins with a discussion of conventional astrology which he refers to as an illness. Yet he rejects the possibility of the star appearing by accident at the same place and time as the great conjunction. God, he says, adapts himself to men and makes use of the rules of astrology, which are in themselves objectionable, in order to inform them of his opinion. Concerning interpretation, Kepler is like a compelled 'beast of burden' which has to put 'its foot in this puddle'. He feels it would be best if, in the presence of the celestial sign, people were to commune with themselves, examine their mistakes and vices, and repent.

There has been no recurrence of a naked-eye supernova since the time of Kepler. The world's modern astronomers await the next one with high anticipation, for Brahe's and Kepler's stars appeared on the historical scene before the telescope did, and they are now ready with extremely powerful and sophisticated instruments. As Clark and Stephenson, in their book *The Historical Supernovae*, say of a new supernova occurring within our galaxy:

> The immediate impact on modern astronomy would undoubtedly be considerable – the long-term impact on civilisation on the planet earth of a particularly nearby supernova could be extremely dramatic . . . the world's astronomers and astrophysicists wait, hoping that during their lifetime they may be privileged to witness one of the Universe's greatest spectacles.

Man's philosophy may change with the ages, but each philosophy cannot avoid turning upwards to the stars in recognition of events which contain the very source and secret of life.

Between 1572 and 1604 there were other reports of new stars, one by David Fabricius in Frisia in 1596 when he noticed an addition to the constellation of the Whale between August and October. A new star was also reported in the Whale in 1638 by the Frisian astronomer Holwarda who wrote an account of its discovery and eventual disappearance. But after sending his text to the printer the star reappeared, and in an appendix he gave the first account of a fluctuating or variable star. This was also Fabricius's star, and in August 1659 Holwarda named it 'Mira' – the 'wonderful' star (Figure 10.21). It was left to Ismael Bullialdus a few years later to recognise that the variation in brightness was regular and reached a maximum every eleven months.

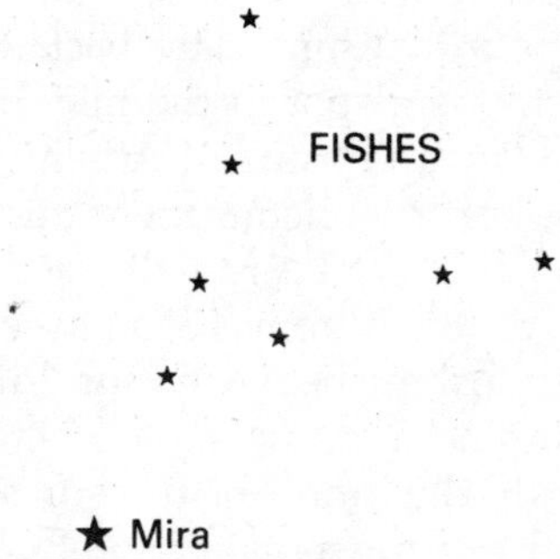

Figure 10.21

Today the astronomer, with the aid of the telescope, identifies about 25,000 variable stars. Besides Mira, a naked-eye example is Algol in Perseus (Figure 10.22) noted by Montanari in

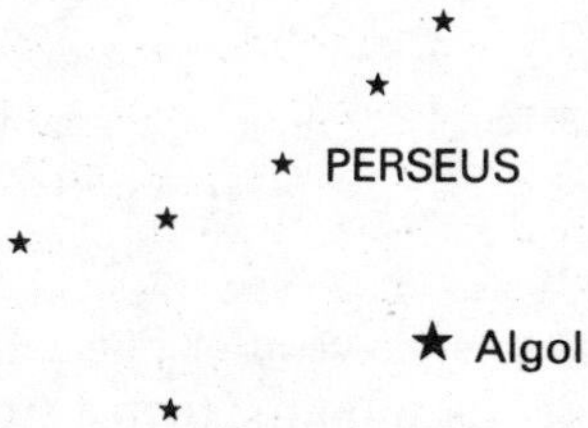

Figure 10.22

November 1670. But the name is said to come from the Arabic 'El Ghoul' which authorities claim means 'changing spirit' or 'mischief-maker'. Thus it has been known as the Demon Star, the Blinking Demon, and (from Ptolemy) 'the bright one of those in the Gorgon's head'. Unlike Mira, Algol does not disappear from naked-eye view when at minimum brightness. Its variation cycle is just under three days. The intervals between maximum brightness of stars can range from hours to years in a regular, semi-regular or irregular manner. In fact, all stars are understood to be variable. A nova is a 'cataclysmic variable'.

Even the sun is not exempt, for it is realised that, with the presence of sunspots, the brightness of the sun fluctuates.

This brings us to our final consideration in the study of phenomena which appear and disappear – or at least alter in their degree of visibility; these appearances, vanishings and changes arising largely out of the objects themselves.

Records of sunspots in China date back to 28 BC and in Greece a reference to a sunspot in the mid-fourth century BC can be attributed to Theophrastus of Athens, a pupil of Aristotle. Seen from the northern hemisphere, these black spots or patches rotate from left to right across the face of the sun. Their line of movement is different from season to season, expressing the changing relationship between sun and earth. They normally appear within bands which are between 10 degrees and 30 degrees north and south of the sun's equator. Figure 10.23 shows the paths of sunspots for an observer looking due south at noon at different times of year. It should be noted that the sun's equator does not lie in the plane of the ecliptic.

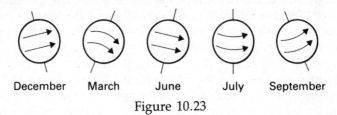

|December|March|June|July|September|

Figure 10.23

If a spot happens to survive after one rotation round the sun (large ones can linger for months) then a strange similarity with the rotation of the moon is revealed. The spots move across the sun's disc at different speeds, depending on their latitude on the sun. They are seen on its face for just under a fortnight at most, and indicate the motion of an outer layer of the sun itself. This establishes an average synodic (seen from the earth) rotation period of 27.27 days – very close to the sidereal period of the moon. Spots nearest the sun's equator move fastest, and at their usual highest or lowest latitude of about 40 degrees they have a synodic period of 29.65 days – roughly that of the moon's synodic period.

In 1843 Heinrich Schwabe discovered that sunspot frequency reaches a maximum about every ten years, later confirmed and corrected to an average of about eleven years when historical records were examined. Then in 1887 Spörer, and later Maunder, reported that there was a prolonged minimum when the sunspot cycle seemed to disappear between 1645 and 1715, though a recent reappraisal of Chinese records casts some doubt on this. However, it is evident that in the last 300 years the cycle has varied between the extremes of eight years and fifteen years. In an average cycle, sunspot activity increases for four years to a maximum then decreases for seven years. After a minimum, spots of a new cycle start at highest and lowest solar latitudes, after which activity gravitates towards the equator.

There have been attempts to link sunspot activity with the weather and with the sidereal period of Jupiter (11.86 years). There is some evidence to show a possible link with global weather. However, it should be added that the oft-cited study of tree-ring widths is found to reveal the domination of local rather than global solar effects. As to a sunspot link with Jupiter, this has not as yet been scientifically established.

Large sunspots are naked-eye phenomena and can be seen when the sun is glimpsed through a dark medium. Two four-teenth-century sightings in Russia were at times of forest fires when smoke would dim the glare of the sun. The sinologist Joseph Needham has suggested that in early China the sun might have been viewed through semi-transparent jade, mica or smoky rock crystal. But possible damage to the retina of the eye forbids any recommendation today that the sun be studied directly, at least without professional advice. A safe method is to shine an image of the sun into a darkened space where the spots can readily be viewed, especially if the image is projected through a telescope or lens. What was probably the first published illustration of a camera obscura is in a book by Rein-erus Gemma-Frisius.** A drawing shows an image of the sun projected through a small hole in the wall of a room onto the opposite wall during the eclipse of 24 January 1544, observed at Louvain. The sun was observed in a similar way by Copernicus, Brahe and Kepler.

Chinese descriptions of sunspots, however they observed them, are charmingly imaginative as well as presumably having some code of size and shape. The first extant record of 28 BC describes the spot as being 'as large as a coin' and others subsequently are described as being 'as large as a melon'; 'like a flying magpie'; 'the shape of a three-legged crow'; 'as large as a hen's egg'; 'as large as a peach'; 'as large as a cup'; 'as large as a plum'; 'as large as chestnuts', etc.

One interesting discovery of modern science is that these black spots we see are, in fact, areas of light. They appear as black because they emit less light than the rest of the sun's surface, and are also cooler. In addition, they are powerful sources of magnetic activity which, in turn, has a connection with other solar activity and the occurrence of aurorae or 'northern lights', which latter phenomena will be discussed in the next chapter. Aurorae occurring at high altitudes in the earth's atmosphere are more numerous during a decrease of sunspot frequency, while aurorae at low altitudes tend to take place mostly near a sunspot maximum.

Such, then, is a survey in this chapter of phenomena which, according to their own mysterious laws, light up in our skies for an unexpected season then pass away. We have concluded with spots and shapes which are also light, but look dark. Here the question of light is relative. However, the observed phenomena are dark and they indicate a dimming of the sun's light, classifying it as a variable star. Also, one cannot predict an individual sunspot. Phenomena from comets to sunspots shake us out of the regular, conventional run of things. We need 'broom stars', 'guest stars', 'constellation children' and 'three-legged crows'.

* Some astronomers doubt it did, though the nucleus transited (undetectably) the sun's face.
** *De Radio Astronomica et Geometrico* (1545).

Chapter 11

Light and the Sky

Between the observer and the drama of the circling stars there lies an intermediate realm of coloured lights which play, sometimes gently, sometimes strongly, onto the celestial stage. They radiate like an aura into the earth's air system, bringing colour to its breathing.

The brightest colours, yellow and red, lie between the poles of day and night, light and darkness, and are strongest at the horizon where sky and earth meet. These are the sunrise and sunset colours, the lords of the horizon which manifest at the time of day when a transition takes place between outer and inner life. Red and yellow, in dynamic fashion, stand at east and west as no meek guardians of these thresholds. These are the colours which speak most, which announce. And every day, every season the colour-face of the sky bears a different expression and different thought behind it concerning Man, nature and the heavens. A theory of colour must include a poetic feeling for every passing mood. Here science and poetry can join hands in a totality of human experience.

It was mentioned earlier that the general impression of the shape of the sky was half of an oblate spheroid – a flattened dome. This impression is pronounced in the twilight of morning or evening, lending greater apparent size to the setting or rising full moon. The oblate celestial spheroid breathes in our experience in the course of twenty-four hours. Generally speaking, the more atmospheric effects there are the more flattened the sky seems. But earth and sky certainly seem to draw close at sunrise and sunset when the strongest colours also flood the scene. With these colours an extraordinary sense of calm and contemplation pervades the mind. Yet sunrise and sunset grow in opposite ways and the joyful awe of a sunrise is quite different to the sobering splendour of a sunset. Turner had no need to label the time of day on his paintings to indicate which end of it he was poetically capturing. The scientist, if he is to be true to nature, must be able to read this script too and include it in his treatise on the effects of light.

In a sunrise or sunset it is important to observe the whole of the sky drama. It is easy to ignore the subtler effects away from east and west within this celestial cranium. If the sky is clear, for example in the morning, a more obvious observation is the extraordinary moment of division between night and day when the waning moon might be seen among the stars on one side of the sky, the light of dawn having already put the stars to flight on the other side. The so-called celestial hemisphere arches above, and left and right (east and west) resolve themselves into two 'cerebral' hemispheres. The north-south meridian marks the cerebral 'longitudinal fissure' and the pole star stands at the back of the cranial vault. The anterior fontanelle lies at the zenith. East and west are the temples. In the course of twenty-four hours each cerebral hemisphere becomes a location for horizon colour – but each in a different way. Just as the two hemispheres of the brain are known to be the seat of different faculties, so the morning and evening sides of the sky have their separate characters. This pictorial description is here related to observations in northern geographical latitudes. To the east and left are the rising, awake elements of practical consciousness, while to the west and right are the setting, inward elements of feeling and imagination.

Phenomena concerning the middle part of the day and of the night we shall deal with later on. For the moment it is necessary to carry further the picture of the illuminated celestial dome when lit from east or west. It is of interest that when the sun is in the process of setting in the west, a 'counter-twilight' develops in the east. In clear sky conditions about twenty minutes after sunset, western and eastern horizon colours approximate those indicated in Figures 11.1 and 11.2. The grey-blue of the counter-twilight is the earth's shadow rising. These eastern colours at sunset have a subtle transparency, altering in blend and hue as one watches and manifesting a delicate 'peach blossom' at one stage. The statement of the sun at one horizon is answered at the other. The purple of the counter-light seems strongest in late summer and autumn.

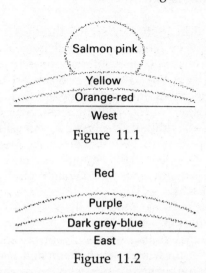

Figure 11.1

Figure 11.2

Above the setting or rising sun a tapering column of light can often be seen stretching vertically into the sky and this is referred to as a sun pillar. Occasionally, before the sun sets, a short column can be seen below the sun. The vertical position of the pillar distinguishes it, in non-equatorial regions, from the rarer phenomenon called zodiacal light. As the name suggests, this faint band of light above the rising or setting sun stretches along the zodiac like a gentle, pointing finger. Most favourable times for observation are when the angle of zodiac and horizon is greatest – i.e., at spring sunsets and autumn sunrises. Conditions must be dark with the the sun well below the horizon, when a rounded pyramid of soft, glowing light not unlike the Milky Way extends into the zodiac.

In summer the sun does not sink far enough below the horizon in mid-northern latitudes for the zodiacal light to show, but at other times the light can sometimes be seen extending thinly round the whole arc of the zodiac, as if manifesting the special life of this important circle of stars. The zodiacal light has been known since antiquity and is also referred to as the 'false dawn'. On October mornings the star Regulus stands within the pyramid of light in the east.

During a dark, clear night a counter-sun can be seen to move from east to west with the zodiac, in exactly the opposite position to the sun. Averted vision from the exact spot reveals this faint patch of light called Gegenschein or counter-glow, first discovered in 1854. It is on the threshold of visibility for the naked eye, yet it cannot be seen with binoculars or telescope as they narrow the field of vision and separate the seeing from wider, darker contrasts. The eye also has an advantage over the camera which can only record the light with difficulty and possible distortion. If the zodiacal light is present also, the Gegenschein lies within it as a broader, intensified area at the anti-solar point. A good time for seeing the Gegenschein is in October when it appears in the dark sky area of the Fishes. Later in the winter it blends into the light of the Milky Way in the Bull and Twins.

Such phenomena as zodiacal light and Gegenschein are essentially extensions of the sun into the dark hemisphere of the sky. The night zodiac is not without the subtle, luminous influence of the sun within which moon and planets wander.

Considering the night sky as a whole, one can speak of the sun's subtle presence over all of it. If one stays under the stars on a moonless night for several hours, it is surprising how well one can see. Even newspaper headlines can be read without difficulty. The stars contribute about one-third of such a dark sky's illumination, zodiacal light about another third, and the rest is due to what is called nightglow or earth light. This delicate and transparent veil of light, occurring many miles high in the atmosphere, is thinnest (faintest) at the zenith and thickest (brightest) about 15 degrees above the horizon. Even in the darkest hours the sun's light still has its effect.

Under clear conditions, the time when the sky is at its deepest blue is at dawn or dusk and in the overhead direction. After the sun has risen, this bluest and darkest daytime point can be found between 95 degrees and 65 degrees from the sun (depending on how high it has risen) along a line of altitude. Again under clear conditions, the blue of the sky is always darker at the zenith than it is along the horizon where it takes on a whitish tint. The complement to this is when the sky is evenly overclouded and then the zenith becomes brighter than the horizon.

This leads us on to other daylight phenomena and to 'extensions' of the sun during the day itself. The simplest is a corona (Latin: crown) or ring round the sun (also referred to as an aureole). Such effects are frequent, but not noticed because of the brightness of the sun. Newton made his well-known

observation of a corona by observing the sun's dimmed reflection in calm water. They can also be seen if the direct line
between eye and sun is screened off by part of a building, etc.

The diameter of the corona is only a few degrees and much
smaller than another daytime phenomenon of this sort – a halo
(Greek: threshing-floor) round the sun. Here the sunlight in the
atmosphere (generally in veils of cirrus clouds) where hexagonal, prismatic ice-crystals form, appears pressed out and
refracted into a ring with a radius reaching from the thumb to
the little finger of one's outstretched hand held at arm's length
– about 22 degrees. Colours range inwards to outwards through
red, yellow, green, white and blue. The sky inside the ring
often seems darker than outside. The 22-degree halo is most
readily seen in April and May.

Under favourable conditions, this halo is accompanied by
other, often more complicated geometrical workings of sunlight
into the water element in the atmosphere through refraction
and reflection in minute crystals. Figure 11.3 is taken from a

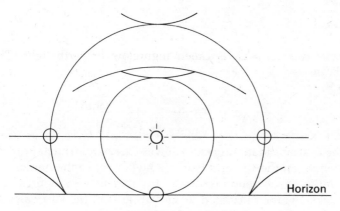

Figure 11.3

drawing made in 1820 by the explorer Sir William Parry during
his voyage in search of a northwest passage. The smaller circle
of light round the sun is the halo previously described, and the
outer circle is the 46-degree halo. The horizontal line of light
passing through the sun is the parhelic circle and at the intersection of this and the smaller halo are two 'mock suns' or 'sun
dogs' or parhelia which can glow quite intensely – red on the
inside, spreading out into yellow and bluish white. The circum-

scribing arc above the small halo is named the upper Parry arc, after the explorer, and the arc tangent to the top of the large halo is the circumzenithal arc – part of a circle with its centre at the zenith.

These are some of the main features of this drawing, but the sun-geometry displayed by these phenomena can become much more complicated and sometimes difficult to explain optically. Also, the set patterns can vary at different appearances, depending on the altitude of the sun, etc. Research still continues, based on the geometry of hexagonal prisms and reflectors. One aspect of this research is the use of the computer to simulate the necessary geometrical and optical conditions which otherwise could not be calculated. This discovers certain geometrical principles involved, one example being found in the Parry arc indicated in Figure 11.3. The upper Parry arc above the 22-degree halo is shown in Figure 11.4. Simulations

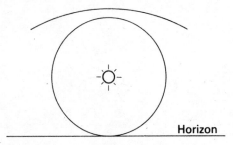

Figure 11.4

of the optical situation reveal a lower Parry arc, which, indeed, is seen when the sun and halo are higher above the horizon or when the phenomenon is viewed from an aircraft. Figures 11.5 to 11.9 show the calculated metamorphosis of the Parry arcs as they are seen at different elevations, the horizontal line representing the horizon. In Figure 11.9 the sun is 70 degrees above the horizon. Haloes represent the geometry of the ice-crystal (parent of the snowflake) and the sun raised to a high degree. Human ingenuity is still struggling to solve these exalted 'theorems'. Writing a skilful technical description of halo phenomena in the *Scientific American* of April 1978, David Lynch ends by saying: 'Halos stir one's mind and soul, since they probe both the physical environment of the cloud and one's awareness and appreciation of the natural world.'

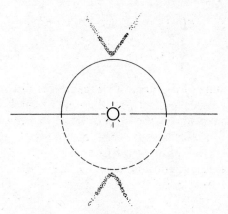

Figure 11.5

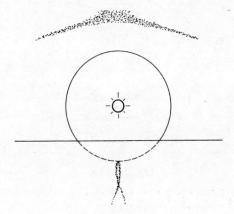

Figure 11.6

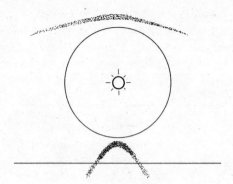

Figure 11.7

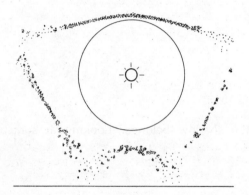

Figure 11.8

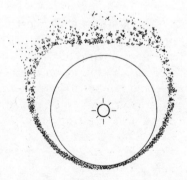

Figure 11.9

Famous representations of complex halo phenomena include a painting by Olaus Petri of a display over Stockholm on 20 April 1535; a drawing entitled 'Seven Suns' by the Danish astronomer Hevelius of the display he saw in Gdansk on 20 February 1661; and a drawing of a display seen over St Petersburg on 18 June 1790, by Tobias Lowitz.

When certain effects combine, like the sun pillar, the parhelic circle and the 22-degree halo, then crosses of light can appear centred on the sun or to either side of it. Also, pillar and halo effects take place at the opposite side of the sky to the sun, centred on the anthelion or anti-solar point which can appear as the intersection of arcs. This anthelion point, when the sun is above the horizon, can form on nearby mist if the observer is standing on a mountain top with the sun behind him. Against the mist is seen the shadow of the observer's head surrounded by a coloured halo. This is known as the 'glory' – or the 'spectre of the Brocken' from its frequent appearance on the highest peak of the Harz Mountains called the Brocken. Diffraction rings appear in the mist and can also be seen from aircraft, round the shadow of the plane on cloud. The spectre of the Brocken round someone's head can only be seen by the person to whom the shadow belongs. Looking at someone else's shadow, one cannot see it. It was on the Brocken on 4 September 1784 that Goethe felt moved to pen the words of Manilius in the visitor's book kept at the top: 'Who could know heaven save by heaven's gift and discover God save one who shares himself in the divine?'

Connected with all these phenomena is the majesty of the rainbow, which would require a chapter to itself if treated in any detail. Needless to say it too can manifest many subtleties including white rainbows, red rainbows, secondary bows, reflection bows, interference bows, etc.

On the night side of the sky the moon can be the source of pillars, coronae, haloes and rainbows. Even Venus, Jupiter and bright stars can appear with faint coronae round them. The moon affords the commonest observations of coronae and 22-degree haloes which show up against thin clouds and dark sky. Moonlight, as said, gives rise to rainbows, though they are rather faint and usually colourless.

The moon corona is often a small band with bluish light on the inside, expanding into white then reddish brown. It can sometimes be surrounded by one to three other coloured rings, extending to a maximum radius from the moon of 13 degrees. Thus on rare occasions a 'four-fold corona' is referred to.

A special but well-known phenomenon in connection with the sun is the green flash or green ray. Just as the sun is setting and the top edge is about to disappear, an emerald light shoots up above it for a moment. The best conditions are a distant sea horizon and a bright sun, not too red. From the rolling deck of a ship it can be seen several times in succession. The Dutch scientist Minnaert prolonged a view of the ray for twenty seconds by running up the slope of an 18-foot dyke. It can also be seen above the rising sun, and during Admiral Byrd's expedition to the Antarctic in 1929 was observed on and off for thirty-five minutes as the sun skimmed an uneven horizon in the slow, 78-degrees south sunrise. However, normal sightings are more common in the tropics than in temperate latitudes.

On one occasion the ray has been seen to change from green into blue and violet in the course of a few seconds. Green colour is also seen to belong to the upper edge of the setting sun and to the top section of the refracted, distorted disc which, at one particular moment, appears to separate from the main body of the sun. The green flash itself has been successfully photographed.

Here is an intermediate colour for a precisely intermediate phenomenon between day and night, when the whole range of other colours are manifest in the sky around it. In the Isle of Man it is called 'living light'. An old Scottish legend (writes Jules Verne) says that anyone who has seen the green ray will never err again in matters of sentiment. A ray of pure thought, perhaps.

The green flash appears frequently above the clear desert horizons of Egypt where, in ancient times, people seem to have believed that the sun is green while travelling beneath the earth at night. A stone pillar from about 2500 BC depicts the sun, either rising or setting, as a semi-circle coloured green below and blue above. Interest in the phenomenon in the West was largely stimulated in 1882 when Jules Verne published a science-fiction novel called Le Rayon Vert which describes a search for the mysterious green ray of the sun. Green ray phenomena have also been observed on rare occasions in connection with the moon, Venus and Jupiter. There is, in addition, an infrequently-observed red ray which has been seen at the underside of the sun as it stands just above the horizon. But the bright, emerald light above the sun is the central phenomenon which flashes persistently into our atmospheric environment and has, in recent times, been shown not to be an activity of the eye or an after-image effect. It is a demonstration of Goethe's Colour Theory where the atmosphere acts as a prism and displaces the image of the sun upwards so that sunlight and sky colours overlap and produce green.

Turning our attention to starlight, we can notice effects which do concern the activity of the eye. For example, red stars, if stared at directly and intently, are seen to brighten. Yet whitish to slightly red stars appear brighter when the gaze is averted than when they are viewed directly. Colour thus lends stars a 'spectrum of behaviour' for the eye, ranging from boldness to shyness. Red stars also seem to scintillate more than white ones. If stared at directly, faint stars can disappear altogether from the observer and reappear when the gaze is slightly averted. This demonstrates that different parts of the eye have different receptivity and it is important to allow them all to come into play. Some of the subtler sky effects and colours are obtained through peripheral vision.

In northern lands and also in the far south, towards the earth's poles, inhabitants are privileged to witness the most vast and colourful of nature's displays in the theatre of the sky – the aurorae or northern lights. These great surfaces of shimmering colour can hang like huge curtains in the sky or stand forth like the shields of giant Norse gods in heroic battle. To live under such phenomena must be to feel intimately the presence of powerful forces in upper regions.

The word 'aurora' comes from the Latin, meaning the goddess of dawn. The metaphor of an aurora borealis (its northern manifestation) was suggested by Gregory of Tours (AD 538–594) and established by Galileo and the French scientist Gassendi. The first report of an 'aurora australis' (in the Antarctic) was that of 20 February 1773, by Captain James Cook.

Aurorae are understood to arise from interaction between the sun's activity and the magnetic field round the earth. As mentioned in the last chapter, there is a rhythmic link between the frequency of aurorae and the sunspot cycle. One of the most prominent displays of northern lights in recent times took place on 11 February 1958, after an intense solar flare at the surface of the sun two days previously. Above the earth's geo-magnetic pole, the aurora forms as a glowing ring which, during high activity, becomes larger and descends to lower latitudes. At such times radio communications on earth are disturbed. Seen technically, the aurora is a fluorescent luminosity which has been described by scientists as something of a neon light within nature ('an electrical discharge phenomenon in low-density air') or similar to a television picture with the upper atmosphere as the screen and the earth's magnetosphere as a cathode-ray tube.

Be that as it may, nature's display is unsurpassed. Two basic forms emerge – ribbons or curtains, and cloud-like patches. At times of high activity there is a development from the former to the latter. The ribbons begin in a 'homogeneous arc' of smoothly-graded luminosity, brightest at the bottom and fading into the night sky at the top. Folds then appear (Figures 11.10 and 11.11)

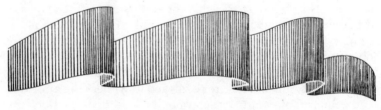

Figure 11.10

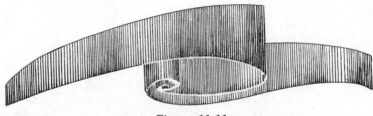

Figure 11.11

and majestically drape across the whole sky like huge curtains or scrolls. These then dissolve and are replaced by separated patches. Such patches appear most often after midnight. In the ribbon forms which develop as far as the stage reached in Figure 11.11, a pink glow appears along the bottom. Most aurorae are green or blue-green, with occasional enhancements of pink and red. However, what has been described are the more commonly-observed forms and colours and there are other, rarer varieties. On infrequent occasions there is a spectacular rose-coloured type.

It is thus fitting to conclude our theme for this chapter with the awesome beauty of one of nature's most arresting sights – through which, incidentally, stars and planets still shine and only add to the splendour. Much of the light of the sky, however, is not so obvious or dramatic. But if the observer knows of its existence he begins to see it. Subtle, unearthly illuminations in a hitherto unconscious realm suddenly become visible.

Such, then, are some of the effects of light in the sky – atmospheric stage lighting for the celestial scene. Without it, astronomy would be a plain affair.

Chapter 12
The Telescope Image and Beyond

This book began with support for an earth-centred astronomy in which appearance, or the immediate experience of the ordinary human being without the intervention of sophisticated instruments, was restored to dignity in order to enter on a fresh view of our knowledge of the sky. For the ordinary person standing under the stars unaided but for his own eyes and a constellation map, this naturally leads to a non-Copernican system. The one adopted here has been largely Tychonic (Chapter 10) with the earth central to the celestial sphere and the sun central to the planets. Now that an observer-based astronomy has been described, it is perhaps in place to relate it to other aspects.

Often the first thing an amateur astronomer is asked by the inquirer into the lore of the night sky is: 'Do you have a telescope?' The answer, of course, is 'Yes.' Then the second inevitable question comes: 'Can I look through it?' The answer again is affirmative, but maybe with the unexpected rider that the stars will then appear as even smaller points of light and it is much more interesting, at first, to look with the naked eye. As the famous amateur astronomer, Leslie Peltier, discoverer of a dozen comets with his telescopes, said 'fortunate is he whose introduction to the skies comes to him through nature's eyes alone and not through any telescope' and stated 'most emphatically' that

> a telescope is not essential to an enjoyment of the stars;
> . . . even without optical aid of any kind one still can
> become an accomplished star-gazer. No one, as yet, has
> ever nearly exhausted all the possibilities of observing
> with the naked eye alone.

But the telescope has come to mean 'astronomy', and in the public mind has taken on a certain magic. And magic it is to peer down the tube and see, moving in the glass, the gliding image of Jupiter and its moons, Saturn and its rings, or crescent Venus. These are incredibly beautiful images which modern invention has brought to the eye of Man. They stretch the mind out into space to gaze wonderingly on other worlds. The tube pushes against the celestial sphere and shatters it.

The moment one glances through a telescope, the ordinary senses retreat and the mind leaps into theory and speculation. Sky phenomena are no longer encompassed by pure observation but for ever stretch beyond it, asking to be interpreted in the rational terms which the optical instrument first lays down. This is an entry into another realm. The universe becomes purely physical. The actors and actresses on the celestial stage are analysed in terms of atoms and molecules. Why did that character suddenly raise her arm and cry out? Because certain muscles contracted, others expanded and there was a vibration caused by compressed air in the region of the larynx. The play is a forgotten myth. The shadows multiply and dance more vigorously at the back of Plato's cave in his allegory (*Republic*, Book 7).

Yet, in themselves, the images in the telescope are a fascinating and wonderful world, just as the larynx is a wonderful organ and yields much of its physical properties to the microscope. The important thing is that the investigator does not lose his sense of wonder and his concept of the whole. But the temptations of the optic tube in the other direction are great.

The word 'telescope' is from the Greek meaning a far view, and is thought to have been devised by the Greek poet and theologian John Demisiani and publicly conferred on the instrument in the presence of Galileo by Prince Frederick Cesi on 14 April 1611, at a banquet in Rome held in Galileo's honour, at which Demisiani was also present. To Galileo, however, it had been simply an 'occhiale' or eye-glass, sometimes also rendered as spy-glass. Milton called it a glazed optic tube.

In May of 1609 when Galileo was forty-five years of age and professor of mathematics at Padua, he heard, while visiting Venice, that a Dutchman had constructed such an instrument. He returned to Padua and within twenty-four hours of his arrival had constructed his own 'by way of pure reasoning' and 'through deep study of the theory of refraction'.

According to the scientist Sir Oliver Lodge, Galileo took an old, small organ pipe and fitted spectacle lenses into either end to construct his first telescope. It magnified objects by three times their diameter. In one end of the tube, probably about 2 feet long and described by Galileo as made of lead, he placed a lens less than 2 inches in diameter which was convex on one side, and in the other an eye-piece which was concave on one side (Figure 12.1). The image he saw was the right way up,

Figure 12.1

unlike later astronomical telescopes which are constructed differently. His first thought for its use was military, as was that of the Dutch who were warring against Philip II of Spain and who tried to keep the invention secret. Galileo returned to Venice with a second version of his instrument. Noblemen and senators climbed the steps of the highest church towers in Venice to see ships heading for the harbour two hours before they were sighted with the naked eye.

Back in Padua, he carefully ground his own lenses and built superior models. The next magnified by eight diameters and was followed by one magnifying by twenty diameters. With this last telescope he observed that the moon had mountains on its surface. His next (fifth) telescope magnified by thirty diameters and he later affectionately called it his 'old discoverer'. On 7 January 1610, he turned it towards Jupiter which was rising fairly high in the east at sunset. The moon was almost full, standing near the planet and shedding its interfering light when Galileo aimed his spy-glass. He saw three tiny points of light near Jupiter and took them to be stars (Figure 12.2). The following night he happened, by mere accident he says, to observe Jupiter again and, to his surprise, the nearby 'stars' had moved (Figure 12.3). During subsequent observations he saw four points of light (e.g. on 13 January as in Figure 12.4) and later announced his discovery of four moons of Jupiter.

The telescope he was then using (his 'old discoverer') was 5½ feet long with a convex lens 2¼ inches in diameter at the

Figure 12.2

Figure 12.3

Figure 12.4

opposite end to the eye-piece (the 'effective' diameter of this convex or 'object' lens, however, was only 1½ inches). This lens, later broken, is now preserved in the Museum of Physics and Natural History in Florence. This would show the moons of Jupiter larger but less clearly than a pair of modern field binoculars. Two of Galileo's telescopes plus the broken objective lens just mentioned were tested in the 1920s in an Italian observatory when Jupiter's moons were again observed through them. They were mounted on telescope equipment which held them steady and moved them in the same direction and speed as the celestial sphere. But Galileo had none of this assistance when he made his observations from, it is believed, either the flower and vegetable garden or window of a large tenement house in Padua. There is no record of a constructed telescope support, and in his writings he simply speaks of fixing the 'tube in some stable place to avoid the trembling of the hand which comes from the throbbing of the arteries and from breathing'. The modern Italian astronomers, after looking through his original instruments, could only admire the sharpness of his eyesight.

Seeing moons (or satellites) move round Jupiter while it, too, moves through space supported the thought that the earth and its moon might do the same and vindicate Copernicus – though none of Galileo's observations proved that the earth was in motion. Further support came with Galileo's discovery in the autumn of 1610 that Venus had phases like the moon which were consistent with it orbiting round the sun. He had also observed in the summer of 1610 that Saturn was 'three-bodied' having a 'star' on each side of it, which stars, however, gradually disappeared over the following two-and-a-half years. In a letter to a patron he wrote: 'Now what can be said of this strange metamorphosis? . . . Has Saturn devoured his children? Or was it indeed an illusion and a fraud?' But the 'children' returned and Galileo further reported that in 1616 he saw Saturn with two 'mitres' or 'ears' on either side instead of round stars. It had to wait until 1656 before the Dutch astronomer Christiaan Huygens, using a long telescope which magnified by a hundred diameters, discovered the secret that 'Saturn is surrounded by a thin flat ring not touching it anywhere, which is oblique to the ecliptic.' This explained the fifteen-year cycle of brightening and dimming, the planet's brightness being enhanced when the rings are seen 'open' from an angle and reflecting more light, and diminished when presented edge-on.

As well as engendering support and admiration, Galileo's discoveries brought powerful opposition, as history knows. The traditional cosmos was being threatened. One opponent said '[T]he satellites of Jupiter are invisible to the naked eye, and therefore can have no influence on the earth, and therefore would be useless, and therefore cannot exist.' Clavius in Rome said of Jupiter's satellites: '[Y]ou must construct a telescope which would first make them and then show them.' Some dissenters refused to look through the telescope at all.

If they had looked through they would have seen those tiny points of light which amazed their discoverer and today thrill the novice astronomer with their strange and serene beauty etched against a sky which is darkened to black velvet by the telescope lens. It is one of the most moving of modern sights. Yet it is so simple. It impresses because it whispers, from the silent spaces, of life out in the universe like our own. Of importance to serious objectors to the telescope's revelations was unease that the celestial world would become included in the physics of the terrestrial.

The eye-piece of a telescope provides a peepshow into a plurality of worlds. If asked, most people today would probably say they believed that there was intelligent life in the universe beyond our solar system. Although many of these same people will not have looked through a telescope, and although the idea of other worlds like ours pre-dated the telescope, the work of this marvellous instrument has nevertheless by its very nature promoted the concept of an inhabited universe. This tendency is furthered by the spectroscope, which analyses the light and colour properties of stars as an indication of physical properties, and radio receivers which detect distant sources which are even invisible to the telescope.

As I write, a newspaper headline has just proclaimed: 'Telescope lifts heaven's veil.' The story is that an orbiting satellite (IRAS), carrying an infra-red telescope which is sensitive to heat and can penetrate inter-stellar dust, has detected 'tiny particles gathered in a giant shell' round the star Vega, which particles 'may one day coalesce into planets'. No planet has yet been detected as a companion to any star except the sun, though there are 'binary' systems of double stars. No 'optical' telescope can resolve the image of a planet on a distant star, yet the hope is that astronomy may one day somehow detect one. It may do, though this is a long way from establishing the existence of extraterrestrial life.

In fact, among scientists the pendulum has begun to swing away from optimism and towards scepticism as regards intelligent life in the universe beyond the earth. This is a recent trend. Only thirty years ago, the Astronomer Royal, Sir Harold Spencer Jones, wrote a book, *Life on Other Worlds*, in which he says that on Mars 'we appear to have direct evidence of life' and 'it is almost certain that there is some form of vegetation on Mars' where 'the question at issue is not whether the canals exist or not. There can be no question about the existence of at least the most conspicuous of them.'

We now know that the Astronomer Royal, and many others with him, were working under an illusion. The illusion was created by the image in the telescope and the fertile imagination of Man. It started last century when, in September 1877, the

Italian astronomer Giovanni Schiaparelli made telescopic observations of Mars during its favourable opposition and noticed
between the 'continents' dark streaks which he called *'canali'* –
Italian for channels or grooves. However, the word was translated as 'canals', and a term for 'artificial' water channels passed
into astronomical literature, leading to intense discussion and
speculation. Schiaparelli later described canals as becoming
double, forming two virtually parallel straight lines principally
in the months preceding and following the melting of the
Martian polar cap which appears in the telescope as a seasonally
expanding and contracting white area.

The idea of artificial canals was followed up strongly in
America by the astronomer Percival Lowell who also saw the
straight, geometrical canal patterns (shown diagrammatically in
Figure 12.5) and concluded that they must have been made by

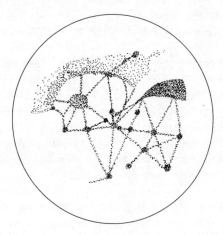

Figure 12.5

intelligent beings struggling to preserve life on a planet which
was drying up by using a great pumping system on a scale far
surpassing any of the works of Man, presupposing an advanced
type of intelligence.

Spencer Jones dismissed the straight lines as optical effects,
but retained the concept of vegetation on the planet. Schiaparelli
himself, though he saw the lines, did not directly adduce the
presence of intelligent Martians, but commented: 'I am very
careful not to combat this supposition, which includes nothing

impossible.'

In 1902 the British astronomer Walter Maunder conducted a Martian canal experiment with a group of boys at Greenwich Hospital School. He hung at the front of the class a small diagram of Mars with surface features shown but without canals. He asked the children to draw what they saw, and those at the back of the room connected some of the smaller markings into lines which looked like the mysterious canals. The French astronomer Flammarion, who supported the idea of Martian inhabitants, heard of the experiment and repeated it with French schoolboys, none of whom drew lines. Perhaps the French children had better eyesight. Perhaps the English children had more imagination. In any case, when spacecraft orbited and mapped Mars at the end of the 1960s, the canals vanished.

Spencer Jones was more cautious concerning the possibility of life outside the solar system. In a more carefully argued conclusion to his book *Life on Other Worlds* he says:

> [W]e are certainly not justified in supposing that the solar
> system is unique. It has somehow come into existence,
> and it is not logical to suppose that other systems could
> not come into existence in a similar way . . . the number
> of planetary systems in the whole universe could be
> considerable, because the number of stars in each of the
> separate stellar universes and the number of these
> universes are both very great.

Spencer Jones does not then fall into the trap of saying that, for the same reason, intelligent life must exist elsewhere in the universe. Such a line of thought is often taken up today, but it is almost equivalent to saying that if a team of monkeys were taught to type and did so for a hundred years (or any considerable length of time) one of them would produce a play by Shakespeare. Instead, Spencer Jones says:

> In any planetary system everything seems to be weighted
> against the possibility of the existence of life . . . Life
> elsewhere in the universe is therefore the exception and
> not the rule . . . We cannot hope ever to have any direct
> information about these remote worlds.

The astronomers Jeans and Eddington both questioned the existence of intelligent life elsewhere in the universe and more recently Sir Bernard Lovell, Professor of Radio Astronomy, has given further support in this direction. In his book *In the Centre of Immensities* he points out that even if the theory of planetary systems elsewhere were correct, then the growth of habitable atmospheres on them needed very special conditions, as evidenced by Mars and Venus probes, and in addition, as in the case of the earth, there was 'almost zero probability' of organic life subsequently developing. Iosif Shklovsky, a leading Soviet astronomer, wrote in the 1980 *Yearbook of Astronomy*:

> If there are no visible signs of highly-developed intelligences elsewhere in the universe, we must conclude that such intelligences do not exist . . . No visiting, super-intelligent beings will arrive to help us sort out our problems on Earth. The responsibility is ours, and ours alone . . . The realization that humanity is a lonely and singular phenomenon in the universe, coupled with an awareness of the vast scale of the cosmos and the frailness and smallness of our beautiful planet Earth, should become a major moral and ethical factor in our thought.

Yet the modern mind persists in populating the universe at distances beyond the resolving power of the telescope, depending on an idea no less fixed than that of Galileo's critics rather than on phenomena.

Sir Harold Spencer Jones also gives voice to a huge and generally held assumption which has followed on from Galileo's first peering through his tube in his Italian garden. He says unequivocally that '[T]hroughout the whole universe there is an essential uniformity in the structure of matter . . . matter obeys the same laws throughout the universe.' This could well be a statement on the level of his Mars observations. The spectroscope only reveals what the spectroscope can reveal and the human mind can only make of it what it knows from experience of the earth. But the universe may be more wonderful than present instruments and understanding allow. To assume that the universe is just like ourselves, only bigger, conditions us to what we think we see and, in this sense, is strangely reminiscent of astrology which stands at the opposite pole ideologically. Traditional astrology and pure astrophysics are the extremes. What is required is a clear-minded but inwardly flexible balance between the two.

At the threshold of outer perception, the realm of imagination begins. Weak telescopes have given us immaterial canals on Mars and even, this century, vegetation on the moon. This imaginative sphere around us will never be pushed back far enough to dispel it. It will always be part of the cosmos whether we like it or not. It invites us to populate the universe with intelligent beings. Ancient religious mythology placed higher intelligences among the stars. Modern scientific mythology places higher intelligences among the stars. The gods have become extraterrestrial civilisations.

There is another point to mention about the telescope which is not usually considered important. Not only does the astronomical telescope narrow down the field of view and the part of the eye used, separating them from the whole, but it also inverts the phenomenon. The number of lenses used is reduced to increase light-gathering properties, which results in the image being seen upside-down; so that when looking at Mars, north is south and east is west. Optically, of course, this is a triviality, but the whole sense of normal orientation is dissolved and left-right, up-down lose importance. It is interesting that Galileo retained normal orientation with his particular arrangement of lenses and when he looked at Jupiter's moons he saw them placed as the naked eye would see them if it could – as in Figure 12.6, and not as the modern astronomical telescope would show the same phenomenon, as in Figure 12.7.

Figure 12.6

Figure 12.7

Galileo's telescopes showed an area of the sky of about half a moon diameter. His drawings of the lunar surface show the moon the right way up, as it is observed with the naked eye, with north at the top and east in the direction of sunrise on earth (Figure 12.8). But later representations made with the inverting telescope turned it upside-down in many maps and books (Figure 12.9). Other maps mix the two axes of orientation (Figures 12.10 and 12.11). This has led to confusion and it is

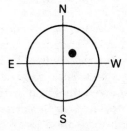

Figure 12.8

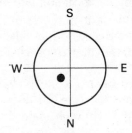

Figure 12.9

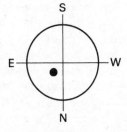

Figure 12.10

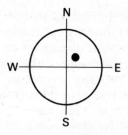

Figure 12.11

sometimes difficult to know, when consulting a moon map, what the directions are. The area designated the 'Sea of Serenity', for example (marked with a dot in the diagrams here), can appear in the east or west. The logic of the orientation in Figure 12.11 is that, viewed thus from earth with the naked eye and north at the top, it is also correct geographically for someone on the moon in relation to an eastern sunrise there, and this layout was preferred by astronauts. In 1961 the International Astronomical Union adopted the latter orientation as official. One remaining complication is that the Mare Orientale (Eastern Sea) now stands in the western hemisphere of the moon's globe. The term 'sea' is a relic of the days when astronomers thought the dark patches on the surface were water.

When Galileo turned his telescope to the fixed stars he found that they did not show up as round discs like the planets but as bright points of light which, he said, had been stripped by the telescope of the extraneous, sparkling rays seen by the naked eye. His spy-glass pierced through what he considered to be unnatural light effects and also confirmed the Milky Way to be a myriad of stars crowded together 'making its nature manifest to the very senses as well as to the intellect'. It also pierced, without knowing it, to the realm of undiscovered planets beyond Saturn. At about 3.45 in the morning on 28 December 1612, he drew the positions of Jupiter and its satellites and added a nearby 'fixed' star which was, in fact, according to recent research, the planet Neptune. No other star visible in his telescope was in that position at the time. Jupiter then occulted Neptune on 4 January 1613, but Galileo saw the planet again on 27 and 28 January, noting at the bottom of his sketch for the 28th that one star (Neptune) had 'seemed further apart' from another the night before. Neptune had just passed its eastern stationary point and had slowly begun to retrograde westwards towards the other star (SAO 119234) in the modern constellation of the Virgin (Figure 12.12).

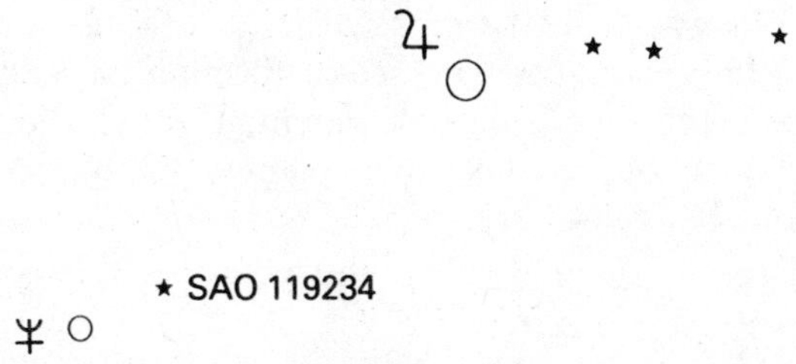

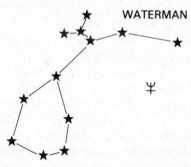

★ SAO 119234

Figure 12.12

Given a few more observations of the moving star, and trust in his first impression that it did in fact move, Galileo would have been the first astronomer in history to discover a planet beyond Saturn. As it was, Neptune then moved all the way round the zodiac and into the constellation of the Waterman before it was eventually seen and identified as a planet by Johann Galle at Berlin Observatory on the night of 23 September 1846 (Figure 12.13). He used calculations, made by the French

WATERMAN

Figure 12.13

astronomer Le Verrier, for the position of the planet. They assumed the planet to exist because of a gravitational disturbance of the orbit of Uranus. On 13 March 1781, Sir William Herschel discovered Uranus, with a telescope in his back garden in Bath (though the planet, at best, is just visible to the naked eye) and Pluto was discovered (using photography) in the same constellation of the Twins early in 1930 by Clyde Tombaugh of the Lowell Observatory, USA (Figure 12.14). The telescope has also winkled out smaller bodies in the solar system, the first minor planet or asteroid between Mars and Jupiter being discovered by the Italian astronomer Piazzi on 1 January 1801. This asteroid was named Ceres and since its discovery thousands of others have been detected, forming an asteroid belt near the plane of the ecliptic and moving through the zodiac in the same direction as the other planets. Only one, Vesta, reaches naked-eye visibility.

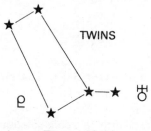

Figure 12.14

The telescope has also reaped its share of misapprehensions and mysteries. A French country doctor and amateur astronomer called Lescarbault, living near Orleans, reported that he had seen what looked like a planet cross the face of the sun on 26 March 1859. The astronomer Le Verrier used this as part of evidence for a planet orbiting the sun within the orbit of Mercury and named it Vulcan – the Roman god of fire. Emperor Napoleon III bestowed the Legion of Honour on Lescarbault. Le Verrier gave Vulcan an orbital period of thirty-three days and predicted that it would transit the sun again on 22 March 1877. The astronomers of the world watched the sun on that day, but no planet appeared and it has never been seen since. There are other explanations as to what Lescarbault saw. Perhaps his telescope deceived him or, it has been suggested, he saw an Apollo group asteroid which passes between sun and earth.

Telescopic study of actual planets revealed for the first time that they have phases like the moon. Mercury or Venus pass through all phases seen from the earth, as shown in plan view in Figure 12.15. Seen side-on from the earth the picture is as in Figure 12.16, with the half phase occurring at greatest elon-

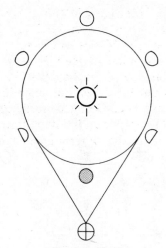

Figure 12.15

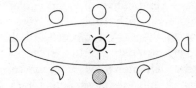

Figure 12.16

gation, where a line of sight from the earth is a tangent to the planet's orbit. The apparent diameter of Venus is eight times greater at inferior conjunction than at superior conjunction, and therefore shows a maximum lit area and is brightest when at crescent phase. Mercury's inferior conjunction is further from earth than Venus', and its superior conjunction is closer to earth than Venus', so its change in apparent diameter is less and the lit-area and brilliancy maximums favour the gibbous or full phases. If we also place the exterior planets such that their centre is the sun rather than the earth (though the telescope cannot distinguish the difference satisfactorily as far as phases are concerned) then it can be seen from Figure 12.17 that they never become less than gibbous. Their least phase is at quadrature and at both superior conjunction and opposition they are full. Even through the telescope the interior and exterior planets display quite different characteristics. The image in the tele-

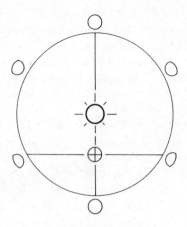

Figure 12.17

scope must be experienced as qualitatively as the naked-eye impression. The interior planets are seen to possess no moons, satellites being the prerogative of exterior planets. Jupiter and Saturn have moons which orbit in a retrograde direction, opposite to that of their other moons. All these phenomena point to specific characteristics of these celestial bodies.

As mentioned in Chapter 8, Mercury passes across the face of the sun in transit, but the telescope is required to observe it. It was first observed thus on 7 November 1631, by Pierre Gassendi, canon of the parish church of Digne in France, following a prediction of the event by Kepler. Mercury's transits take place in November or May, and the November events are more than twice as frequent as those in May. However, May transits are generally slower as the planet is near aphelion and moving less rapidly than in November when it is near perihelion. Although there are no transits of Venus in the twentieth century, there are fourteen of Mercury in this period. The final three this century are on 13 November 1986, 6 November 1993 and 15 November 1999. Returns of Mercury's transits are almost certain in periods of forty-six years, though account has sometimes to be taken of the calendar's leap years.

Telescopic observation of Mercury led to a remarkable modern illusion concerning the rotation of the planet on its axis. Because Mercury is best observed when the ecliptic stands at a steep angle to the horizon on autumn mornings and spring evenings, astronomers most often observed the planet at these times. This meant that three synodic periods of Mercury (each about 116 days) occurred between, say, two favourable evening appearances. At these appearances the planet showed more or less the same markings, so it was assumed that it always turned the same face to the sun, like our moon. This was announced by Schiaparelli in Italy in 1891 and supported in France by Antoniadi. It was not until 1965 that radio astronomy, carried out at an observatory in Puerto Rico, discovered that the situation was otherwise. Radar echoes from the planet establish an axial rotation relative to the stars of 58.6 days. This means that six axial rotations will be completed in about three synodic periods, thus leading to the deception. The surprising aspect is that there appears to be no ordinary reason for this coincidence between the rotational movement of one planet and its orbital relation to another. In addition, a day on Mercury (176 earth days) is longer than the planet's year – in fact, the day is exactly two years long.

An unexpected discovery by radar about Venus was that its solid surface beneath cloud cover rotated clockwise seen from its north pole – the opposite way to normal. It does this in relation to the stars in just over eight earth months. The Venusian year is 7.6 earth months and in that time the planet experiences just under two of its 'days' – the opposite situation to Mercury. It is remarkable that, while Mercury turns its same face to the earth every three synodic periods, Venus does so after every synodic period.

Beyond the planets the telescope forms images of deeper space, yet the stars still do not yield up a physical nature to the inquiring lens. They remain points of light, though sharper and smaller (i.e., without irradiation) than to the naked eye. To vision, the stars reach us not as physical entities but as qualities of concentrated light.

In 1610 Nicholas Pieresc, using a telescope given to him by his friend Galileo, discovered a nebulous glimmer round a star in Orion's sword. It is now known as the great nebula in Orion, lying within our own galaxy, and is understood to consist of gaseous substance and stars. On 15 December 1612, the telescope of Simon Marius of Kulmbach was aimed at a misty, naked-eye patch of light in the constellation of Andromeda and he described it as like 'the light of a burning candle, shining through translucent horn, when seen at night from afar'. The optic tube was probing what now appears as a whole galaxy of stars outside our own: another world. Marius's candle flame is today called the spiral galaxy in Andromeda.

At this stage, the immensity which appears to stand in the telescope eye-piece brings the mind to the borders of incredulity. No longer can the whole be honestly grasped. The detail is overwhelming. One can count it a blessing that the unaided eye cannot perceive such things, otherwise the celestial realm would impose itself mercilessly on the observer from all directions, imprisoning him. As it is, we only perceive these things along a narrow shaft of vision at a time and, if we keep a sense of orientation and selfhood, we seem to be looking into the very inner organs of our cosmos when we glimpse the glowing forms of the galaxies and nebulae.

The telescope has not brought these distant worlds nearer to us, but it has spread out the image of what we are already able to see, to allow us to detect more detail. It is the eye which determines what, and how much, we see in the telescope. The telescopic universe already lives in the eye in this sense. Further, if the eye is conditioned to earth existence, then it sees what is appropriate to it. The same applies to other methods of detection other than the telescope, which detect what lies within their specific realms. Likewise we cannot come to appreciate the music of Beethoven unless something of the spirit of the music already lives within us. The eye lives in the universe but the universe lives in the eye.

In using the telescope which, with the microscope, is one of the most important scientific instruments of modern times, we must keep our sense of humanity and proportion in the light of its fascinating revelations. We must not become like laboratory technicians analysing a painting by Raphael in terms of pigment, colour grading, chemical formulae, etc., and not seeing the painting or the painter. The telescope examines the letters of the starry script. The unaided eye sees the script. The imagination reads it.

Resolving a percept into an image is as active and imaginative as forming a mythological picture. Both are part of the activity permeating the eye with consciousness. This brings us back to the discussion of optics, which is integral to astronomy, towards the beginning of this book (Chapter 2). The experience of seeing takes place in space in a living unity of image, object and outgoing thought. In his book *Visual Thinking*, Rudolf Arnheim says:

> In looking at an object we reach out for it. With an
> invisible finger we move through the space around us,
> go out to the distant places where things are found, touch
> them, catch them, scan their surfaces, trace their
> borders, explore their texture. It is an eminently active
> occupation.

Astronomy, in addition to rather than despite modern technology, must become an 'eminently active occupation' in terms
of observation, thinking and imagination if it is to retain and
develop any real connection with the human being. Much time
has passed since the Babylonians used the star sign ✳ for a
god or lord in their writings or since theatre audiences properly
understood Lorenzo's words to Jessica as they sat together in
a moonlit garden in *The Merchant of Venice* (Act V, Scene 1):

> *There's not the smallest orb which thou behold'st*
> *But in his motion like an angel sings,*
> *Still quiring to the young-ey'd cherubins . . .*

Here Shakespeare connects the cherubim with the sphere of
fixed stars, a tradition handed down from Dionysius the Areopagite and St Paul about the ordering and naming of the heavenly spheres. Nine hierarchies were depicted beyond the earth,
from angels to seraphim, concordant with the nine spheres
from moon to primum mobile – or sphere of 'first moving'
which stood beyond the fixed stars (Figure 12.18). Each sphere
had its specific quality and movement and the system is strange
to the observer of the sky today. Yet, familiar to the naked-eye
observer is a qualitative differentiation among planets and stars.
A careful and methodical study reveals a definite pattern of sky
lore not often realised and which, in a purely phenomenological
way, builds up a many-tiered picture of the immediate effect
on our experience of celestial movements and appearances. The
behaviour of Mars and its satellites can certainly be character-
ised, as can be seen from the present book, by the Greek word
'*Dynameis*' (Figure 12.18).

What, then, lies beyond the image in the telescope and its
technical successors? Human thinking, imagination, and their
vehicle the eye. Today it takes an effort to perceive actively and

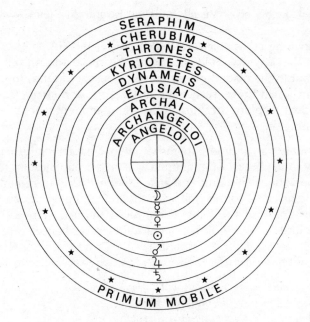

Figure 12.18

unaided as in the days prior to the great discoveries of Galileo. We should use the telescope knowing this and refresh the eye with wide, peripheral views and grow as intimate as we can with the face of the whole sky. The eye has three areas of seeing – central, intermediate and peripheral – as has the mind in understanding.

But since the time of Galileo no need is felt for another revolution in seeing or an augmentation of the spy-glass approach. The situation is excellently put by Jung in his book, *Synchronicity: An Acausal Connecting Principle*:

> When Galileo [he says] discovered the moons of Jupiter with his telescope he immediately came into head-on collision with the prejudices of his learned contemporaries. Nobody knew what a telescope was and what it could do. Never before had anyone talked of the moons of Jupiter. Naturally every age thinks that all ages before it were prejudiced, and today we think this more than ever and are just as wrong as all previous ages that thought so. How often have we not seen the

truth condemned! It is sad but unfortunately true that man learns nothing from history. This melancholy fact will present us with the greatest difficulties as soon as we set about collecting empirical material that would throw a little light on this dark subject, for we shall be quite certain to find it where all the authorities have assured us that nothing is to be found.

There is plenty to be found if we have the eyes to see it. As seeing is active, what we see is equally as much a reflection of the human being as of nature. This is increasingly being recognised today and the Galilean pendulum, so to speak, is swinging back.

I hope that the foregoing chapters will have provided the reader with a description of observational astronomy which lends itself to further precise investigation into the living relationship of Man and cosmos. It has not been my task to write a fanciful description or one which labours 'connections' in an artificial way. Instead, some support has been given for a simple and direct experience of nature – an experience which at best can become transparent, and thus provide a sense organ for what lies behind the phenomena. If this book succeeds in encouraging a few eyes to turn to the heavens afresh, without prejudice, and to wonder and learn about the phenomena of stars and planets as they are seen, then it will have achieved much. For what things are, as they are directly experienced, is the effective environment and contains all of life's sway – a script waiting to be read. Perhaps it is fitting to end by completing the sentence which Lorenzo spoke to Jessica (earlier referred to) and share his concluding thought on Man's usual condition with regard to the voices of the stars:

Look how the floor of heaven
Is thick inlaid with patines of bright gold;
There's not the smallest orb which thou behold'st
But in his motion like an angel sings,
Still quiring to the young-ey'd cherubins;
Such harmony is in immortal souls,
But whilst this muddy vesture of decay
Doth grossly close it in, we cannot hear it.

Appendix 1

'The Earth Does Not Move' – Extracts from a Text by the Philosopher Edmund Husserl

The following extracts are taken from a short, fragmentary text by the German philosopher Edmund Husserl (1859–1938). On the envelope containing the manuscript was written – 'Over-throw of the Copernican system in the usual ideological interpretation. The original-ark earth does not move.'

The extracts are reproduced for their relevance to the subject of geocentricism and for the fact that they come from the pen of a modern thinker, rather than for any particular adherence to Husserl's philosophy or any attempt to represent his entire thinking on the question at hand. I leave their aphoristic sentences to the reader as food for thought as they may be considered to have relevance to our theme in the area where philosophy and science meet. For instance, they provoke the question as to what movement really is, and the need to differ-entiate between kinds of movement and rest.

Husserl's language is peculiarly his own and the reader is referred to the German original entitled 'Grundlegende Unter-suchungen zum Phänomenologischen Ursprung der Raumlich-keit der Natur' in *Philosopical Essays in Memory of Edmund Husserl* edited by Marvin Farber (Harvard University Press, 1940). An English translation was co-published in 1981 by the University of Notre Dame Press and the Harvester Press in *Husserl: Shorter Works* under the title 'Foundational Investigations of the Phenomenological Origin of the Spatiality of Nature'. In his introduction to the latter, the translator Fred Kersten explains that such a work was part of what Husserl described as his 'meditations'. He quotes Alfred Schutz, an editor of Husserl's works, as saying of these meditations: 'These manuscripts of Husserl should not be considered as papers, not even as rough drafts of future literary works, but rather as a philosophical diary, a scrapbook of his thought.'

Kersten also quotes Husserl himself in this context: 'My work is not that of building but of digging, of digging in that which is most obscure and of uncovering problems that have not been seen or if seen have not been solved.'

In Husserl's fragments on the immobility of the earth I find the seed of a new thought which struggles to realise a truth of human experience despite the overwhelming weight of current scientific teaching concerning the purely rational place of the earth in the cosmos. Husserl's musings appear, at first, obscure and idiosyncratic. But, if worked over, they hold together as an organic form in thought and can be grown into the flower which strives upwards in the dark then eventually cracks open the brick. For want of a more pertinent example from modern philosophy, they are quoted here to serve their practical consequences – a phenomenological astronomy; and perhaps contribute to an understanding of Rudolf Steiner's remark* that a time will come when the Ptolemaic system will again be correct.

Extracts from the aforementioned text by Edmund Husserl

The earth itself, in the original idea of its form, does not move and is not at rest; only in reference to it do rest and movement first have a meaning.

But, as earth has become a world body in the unlimited manifoldness of surrounding bodies, rest and movement lose their absoluteness. Movement and rest become necessarily relative. And if there could be a controversy about that, it could only be because the modern-day apperception of the world as world of the infinite Copernican horizon has not become, out of a really followed-through world view, a verified apperceived world.

The apperceptive process has occurred in a way, but it has remained only as an indication for a verifying approach, instead of itself being construed as the ultimate proof.

. . . all proof has its subjective point of origin and

ultimate anchorage in the ego, the proving ego. The proof of the new 'world idea', produced by a modification of interpretations, has its first footing and essence in my field of perception and the orientated representation of that part of the world which has my body as the central body among others . . .

A movement is necessarily relative when it is experienced in reference to a 'ground body' which is experienced as being at rest and with which my body is in union.

The relative ground body is itself relatively at rest and relatively in movement in relation to the surface of the earth which is not experienced, really originally experienced, as a body.

As long as I do not have a conception of a new ground, one from which, in coherent movement, the earth can have meaning as a self-enclosed body in movement and rest, and as long as I do not achieve a conception of an interchange of grounds, and through that a conception of both grounds becoming bodies, the earth is a ground but not a body. The earth does not move – I still say perhaps, it is at rest, but that can only say that each part of the earth that I or others separate off, or that separates itself off, that is at rest or moves, is a body. The earth is a totality, whose parts – when they are thought of on their own, as they can be, as being separated, separable – are bodies, but as a 'totality' it is not a body.

But how can it (the earth) move as a 'whole', how is that conceivable? Not as if it could be firmly grasped – for that the 'ground' is wanting. Has movement, hence bodilyness a meaning for it? Is its place in universal space really a 'place' for it?

My body: in primordial experience it has no outer movement and no rest, only inner movement and inner rest, unlike the external bodies.

I have no movement; whether I stand still or move, I

have my body as centre and around me bodies at rest or in movement and an immobile ground.

But also the ground, on which my body moves or does not move, is not experienced like a body which as a whole can move or not move.

But for all of us the earth is a ground and not a body in the full sense.

'I could fly so high that the earth would appear to be a sphere.' The earth could also be so small that I could explore all of it and indirectly come to a conception of it as a sphere. I thus discover that it is a big spherical body. But that is the quesion, whether and how I would come to the conception of bodilyness, in the sense that the earth would be astronomically one body among others, among the celestial bodies.

The difficulty repeats itself in regard to the stars. In order to be able to grasp them indirectly as 'experienceable' bodies, I must already be a man on the earth as the ground from which my experience originates. Perhaps one could say: the difficulty would not arise if I and we could fly and had two earths as ground-bodies, being able to move between the two by flight. In that way one body would be ground for the other. But what does it mean, two earths? Two parts of one earth with one humanity. Both together would become one ground and each would at the same time be a body for the other. They would have around them a common space, in which each would have a possibly moveable position, but the movement would be always relative one to the other and not relative to the synthetic ground of their togetherness. The positions of all bodies would have this relativity, which would, in regard to rest and movement, result in the question: in relation to which of the two ground bodies?

The stars are hypothetical bodies in a certain as-if sense, and so the hypothesis that they are places of domicile in the sense of being reachable is of a singular type.

Perhaps on the plane of phenomenology it is so, that the

calculations and mathematical theories of the Copernican astrophysics and thus the whole of physics have, within their limits, validity – but it is also the question, whether a purely physical biology – which is biology by virtue of the fact that it is physical – can have sense and validity.

All animals, all living beings, everything which exists at all have validity of being only from the point of view of my constitutive genesis, and this 'earthly' genesis is at the beginning of everything.

There is only one humanity and one earth – all fragments which separate themselves off or have ever done so belong to it. But if that is true, need we say with Galileo: par si muove? And not the opposite, that it does not move? Certainly not in that way that it is at rest in space although it could move, rather as we have tried to represent it above: it is the ark which first enables the meaning of all movement, and all rest, as a mode of movement. Its rest however is not a mode of movement.

By way of further clarification I here add a footnote from Vincent Descombes's book, *Modern French Philosophy*, which comments on this Husserl text, referring to it as 'The Earth does not Revolve'.

It is certainly a striking illustration of the phenomenological step which marks a return to the *lived world* as being at the origin of everything we know, and even the origin, *die Ur-Arche*. At first sight we tend to think that the question of the earth's motion must be decided by astronomy, i.e. by a science dealing with the planet as one celestial *object* among others. Since astronomers have adopted the Copernican solution, we *live* in one world in which we both see and say that 'the sun rises', and we *think* in another, where we know that the earth revolves around the sun. Conflict exists between the lived world (*Lebenswelt*) and the known world, between the *percipio* and the *cogito*. Phenomenology invites us to resolve this conflict by ceasing to identify the real with the objective, and the lived-through with the apparent. It sets out to show

how the lived world lies at the origin of the known, or the objective world. And if the lived world lies at the origin of the *true world*, it must, in its way, be more truthful than the true one.

Science deals with the earth as an object and ascribes to it a movement in space. But this science was born *on the earth*, and it was *here* on this earth that it provided 'objective' definitions of motion, rest, space, and objectivity in general. Scientists' statements, for example the Copernican statement, take their meaning from experiences acquired here. The *here*, which is the place of this first experience, is not therefore a place in space, since it is the place of origin of the very notion of space.

* Page 77, lecture 6, of the cycle 'The Spiritual Hierarchies and their Reflection in the Physical World', Dusseldorf, 12–18 April 1909 (Anthroposophic Press, New York, 1970).

Appendix 2
Calendar of Astronomical Events

1985
Apr 22   Moon occults Mars (seen from Europe)
May 4   Total lunar eclipse
May 15   Spring equinox on N. Mars
Sep 4   Close conjunction of Mercury and Mars
Oct 28   Total lunar eclipse
Nov 29   Summer solstice on N. Mars
Dec 8   Moon occults Mars (seen from S.E. USA)

1986
Triple conjunction of Saturn and Antares
Feb 9   Halley's Comet at perihelion
Apr 24   Total lunar eclipse
Jun 1   Autumn equinox on N. Mars
Oct 17   Total lunar eclipse
Oct 25   Winter solstice on N. Mars
Nov 13   Mercury transits the sun

1987
Mar 29   Moon occults Jupiter (seen from Europe)
Apr 1   Spring equinox on N. Mars
Apr 25   Moon occults Venus (seen from USA)
Jun 11   Close approach of Mercury and Mars
Aug 24   Moon occults Mercury (seen from USA)
Sep 22   Moon occults Mars (seen from Europe)
Oct 17   Summer solstice on N. Mars

1988
Apr 18   Autumn equinox on N. Mars
Sep 11   Winter solstice on N. Mars
Sep 28   Bright opposition of Mars
Oct 7   Moon occults Venus (seen from Europe)

1989
Feb 16   Spring equinox on N. Mars
Feb 20   Total lunar eclipse
Jul 5   Moon occults Mars (seen from Europe)
Aug 5   Close conjunction of Mercury and Mars
Aug 17   Total lunar eclipse
Sep 3   Summer solstice on N. Mars

1990
Feb 9   Total lunar eclipse
Mar 6   Autumn equinox on N. Mars
Mar 22   Moon occults Mars (seen from USA)
Jul 21   Moon occults Jupiter (seen from Europe)
Jul 30   Winter solstice on N. Mars
Aug 12   Close conjunction of Venus and Jupiter
Aug 18   Moon occults Jupiter (seen from Europe and USA)
Oct 16   Close conjunction of Mercury and Venus

1990/1
Triple conjunction of Mars and Aldebaran

1991
Jan 4   Spring equinox on N. Mars
Feb 12   Moon occults Saturn (seen from USA)
Mar 12   Moon occults Saturn (seen from Europe)
Mar 22   Moon occults Mars (seen from Europe and USA)
Jul 15   Close conjunction of Mercury and Jupiter
Jul 22   Summer solstice on N. Mars
Aug 11   Moon occults Mercury (seen from Europe)
Sep 10   Close conjunction of Mercury and Jupiter

1992
Jan 22   Autumn equinox on N. Mars
Jun 1   Moon occults Mercury (seen from Europe)
Jun 16   Winter solstice on N. Mars
Oct 27   Moon occults Mercury (seen from USA)
Nov 21   Spring equinox on N. Mars
Dec 9   Total lunar eclipse

1993
Apr 19   Moon occults Venus (seen from USA)
Jun 4   Total lunar eclipse
Jun 8   Summer solstice on N. Mars
Nov 6   Mercury transits the sun
Nov 29   Total lunar eclipse
Dec 8   Autumn equinox on N. Mars

1994
Feb 14   Close conjunction of Venus and Saturn
May 4   Winter solstice on N. Mars
May 10   Annular solar eclipse (seen from USA)
Oct 9   Spring equinox on N. Mars

1994/5
Triple conjunction of Mars and Regulus

1995
Triple conjunction of Jupiter and Antares
April 26 Summer solstice on N. Mars
May 21   Saturn's rings edge-on to earth

May 27	Moon occults Venus (seen from Europe)
Jun 26	Moon occults Mercury (seen from N.E. Europe)
Aug 11	Saturn's rings edge-on to earth
Oct 26	Autumn equinox on N. Mars

1996

Feb 11	Saturn's rings edge-on to the earth
Mar 21	Winter solstice on N. Mars
Apr 4	Total lunar eclipse
Apr 17	Moon occults Mars (seen from S.E. Europe)
Jul 12	Moon occults Venus (seen from Europe)
Aug 26	Spring equinox on N. Mars
Sep 27	Total lunar eclipse
Dec 22	Close approach of Mercury and Jupiter

1997

Mar 13	Summer solstice on N. Mars
May 4	Moon occults Saturn (seen from Britain and USA)
Jul 27	Close approach of Mercury and Venus
Aug 5	Moon occults Mercury (seen from Britain and N. Canada)
Sep 12	Autumn equinox on N. Mars
Sep 16	Total lunar eclipse
Sep 18	Moon occults Saturn (seen from USA)
Nov 12	Moon occults Saturn (seen from Europe)
Dec 9	Moon occults Saturn (seen from USA)

1998

Feb 6	Winter solstice on N. Mars
Mar 24	Moon occults Venus (seen from USA)
Mar 26	Moon occults Jupiter (seen from Europe and USA)
Jul 14	Spring equinox on N. Mars
Nov 13	Moon occults Mars (seen from USA)

1999

Jan 29	Summer solstice on N. Mars
Mar 6	Close approach of Mercury and Jupiter
Jul 31	Autumn equinox on N. Mars
Aug 10	Moon occults Mercury (seen from USA)
Aug 11	Total solar eclipse (seen from Britain)
Nov 15	Mercury transits the sun
Nov 18	Possible Leonid meteor storm
Dec 12	Moon occults Mars (seen from Europe)
Dec 25	Winter solstice on N. Mars

2000

Jan 21	Total lunar eclipse
May 5	Grouping within 26 degrees of sun, moon, Mercury, Venus, Mars, Jupiter and Saturn
May 17	Close conjunction of Venus and Jupiter
May 31	Conjunction of Jupiter and Saturn
May 31	Spring equinox on N. Mars

Jul 16   Total lunar eclipse
Jul 29   Moon occults Mercury (seen from Europe and Canada)
Aug 1   Moon occults Venus (seen from USA)
Aug 10   Close conjunction of Mercury and Mars
Aug 28   Moon occults Mars (seen from Europe)
Dec 16   Summer solstice on N. Mars

2000/1
Triple conjunction of Jupiter and Aldebaran

2001
Jan 9   Total lunar eclipse
Jun 17   Autumn equinox on N. Mars
Oct 29   Close approach of Mercury and Venus
Nov 4   Close approach of Mercury and Venus
Nov 11   Winter solstice on N. Mars

2001/2
Triple conjunction of Saturn and Aldebaran

2002
Apr 18   Spring equinox on N. Mars
Oct 10   Close approach of Mercury and Mars
Nov 3   Summer solstice on N. Mars
Dec 6   Close approach of Venus and Mars

2003
May 5   Autumn equinox on N. Mars
May 7   Mercury transits the sun
May 16   Total lunar eclipse
Aug 28   Bright opposition of Mars
Sep 29   Winter solstice on N. Mars
Nov 9   Total lunar eclipse

2004
Mar 5   Spring equinox on N. Mars
May 4   Total lunar eclipse
Jun 8   Venus transits the sun
Sep 20   Summer solstice on N. Mars
Oct 28   Total lunar eclipse

2005
Mar 22   Autumn equinox on N. Mars
Jun 27   Close conjunction of Mercury and Venus
Aug 16   Winter solstice on N. Mars
Oct 3   Annular solar eclipse (seen from Spain)

2006
Jan 21   Spring equinox on N. Mars
Aug 8   Summer solstice on N. Mars
Nov 8   Mercury transits the sun

2007
Feb 7   Autumn equinox on N. Mars
Mar 3   Total lunar eclipse
Jul 4   Winter solstice on N. Mars
Aug 28   Total lunar eclipse
Dec 9   Spring equinox on N. Mars

2008
Feb 21   Total lunar eclipse
Jun 25   Summer solstice on N. Mars
Dec 25   Autumn equinox on N. Mars

2012
Jun 6   Venus transits the sun

2040
Sept   Grouping within 29 degrees of sun, moon, Mercury,
         Venus, Mars, Jupiter and Saturn

2061
Jul 29   Halley's Comet due to return to perihelion

2065
Nov 22   Venus occults Jupiter

2084
Nov 10   Earth transits sun's disc with respect to Mars

2100
North celestial pole at its closest to the Pole Star
Nov   Grouping within 29 degrees of sun, moon, Mercury,
         Venus, Mars, Jupiter and Saturn

2123
Triple conjunction of Mars and Jupiter
Sept 14 Venus occults Jupiter

2148/9
Triple conjunction of Mars and Saturn

2163
Nov   Earth transits the sun's disc with respect to Mars
14–15

2368
Spring equinox point enters the stars of the Waterman (equal zodiac division system)

Appendix 3
Astronomical Symbols and Terms

Tropical zodiac of equal 'signs' measured from the spring equinox point:

♈	Aries	♎	Libra
♉	Taurus	♏	Scorpio
♊	Gemini	♐	Sagittarius
♋	Cancer	♑	Capricorn
♌	Leo	♒	Aquarius
♍	Virgo	♓	Pisces

Sidereal zodiac of constellations measured in relation to fixed stars (as designated in this book):

Ram	Scales
Bull	Scorpion
Twins	Archer
Crab	Goat
Lion	Waterman
Virgin	Fishes

Planets:

☽	moon	♄	Saturn
☿	Mercury	♅	Uranus
♀	Venus	♆	Neptune
☉	sun	♇	Pluto
♂	Mars	⊕	Earth
♃	Jupiter		

☌	conjunction	☊	ascending node
☍	opposition	☋	descending node

'Occultation' is when the moon or a planet obscures another celestial object, like a star or a planet. A solar eclipse is, strictly, an occultation.

221

Appendix 4

New Stars of Brahe and Kepler

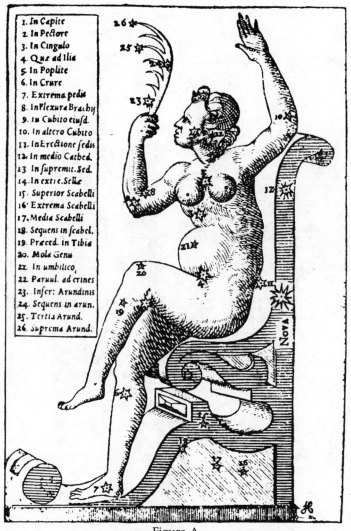

1. *In Capite*	
2. *In Pectore*	
3. *In Cingulo*	
4. *Quæ ad Ilia*	
5. *In Poplite*	
6. *In Crure*	
7. *Extrema pedis*	
8. *In Flexura Brachij*	
9. *In Cubito eiusd.*	
10. *In altero Cubito*	
11. *In Erectione sedis*	
12. *In medio Cathed.*	
13. *In supremit. Sed.*	
14. *In extre. Sella*	
15. *Superior Scabelli*	
16. *Extrema Scabelli*	
17. *Media Scabelli*	
18. *Sequens in scabel.*	
19. *Præced. in Tibia*	
20. *Mola Genu*	
21. *In umbilico,*	
22. *Paruul. ad crines*	
23. *Infer: Arundinis*	
24. *Sequens in arun.*	
25. *Tertia Arund.*	
26. *Suprema Arund.*	

Figure A

The New Star of 1572 (marked 'Nova') depicted in the constellation
of Cassiopeia (seen from outside the celestial sphere, with the star
pattern in reverse to a geocentric view) in Tycho Brahe's *Astronomiae
Instauratae Progymnasmata* (Prague, 1602). See pages 135, 161–6
above.

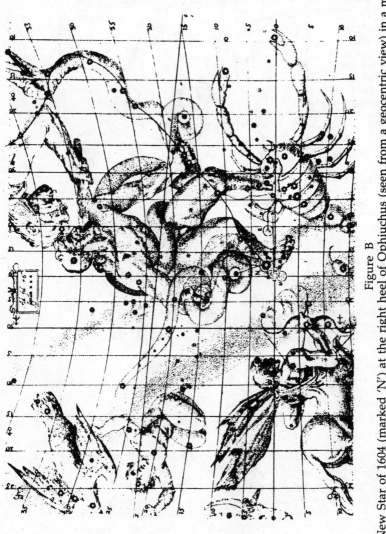

Figure B

The New Star of 1604 (marked 'N') at the right heel of Ophiuchus (seen from a geocentric view) in a map in Kepler's *De Stella Nova in Pede Serpentarii* (Prague, 1606). Near the New Star, changing positions of Mars, Jupiter and Saturn are shown in circles. See pages 167–8 above.

Selected Bibliography

ABETTI, G., *The History of Astronomy*, London, Sidgwick & Jackson, 1954.

AKASOFU, S., 'The Aurora: New Light on an Old Subject', *Sky and Telescope*, 1982, Vol. 64, No. 6, pp. 534–7.

ALLEN, R. H., *Star Names: Their Lore and Meaning*, New York, Dover, 1963.

ANTONIADI, E. M., *The Planet Mercury*, Shaldon, Devon, Keith Reid, 1974.

AQUINAS, T., *The Division and Methods of the Sciences*, Toronto, Pontifical Institute of Medieval Studies, 1963.

ARATUS, *The Phaenomena*, Cambridge, Mass., Harvard University Press, 1977.

ARNHEIM, R., *Visual Thinking*, London, Faber & Faber, 1970.

ASIMOV, I,. *The Tragedy of the Moon*, London, Abelard-Schumann, 1974.

BARFIELD, O., *Saving the Appearances*, London, Faber & Faber, 1957.

BARLOW C. W. C. and BRYAN, G. H., *Elementary Mathematical Astronomy*, Slough, University Tutorial Press, 1956.

BERGH, G. VAN DEN, *The Universe in Space and Time*, London, Scientific Book Club, undated.

BRANDT, J. C. (ed.), 'Comets', *Readings from Scientific American*, San Francisco, California, Freeman, 1981.

BREWER, B., *Eclipse*, Seattle, Earth View, 1979

BROWN, P. L., *Comets, Meteorites and Men*, London, Robert Hale, 1973.

CARSON, R., *The Sea Around Us*, London, Readers Union, 1953.

CHAMBERS, G., *The Story of Eclipses*, Sevenoaks, Kent, Hodder & Stoughton, 1902.

CLARK, D. H. and STEPHENSON, F. R., *The Historical Supernovae*, Oxford, Pergamon Press, 1977.

COLERIDGE, S. T., *Biographia Literaria*, London, Dent, 1971.

COPERNICUS, N., *On the Revolutions of the Heavenly Spheres*, Newton Abbot, Devon, David & Charles, 1976.

CUMONT, F., *Astrology and Religion among the Greeks and Romans*, New York, Dover, 1960.

DAETWYLER, J. J., 'Nautilus and the Dynamics of the Moon', *Journal of Anthroposophic Medicine*, 1981, No. 1.

DAVIDSON, N., 'The Mysterious Inner Planets', *Anthroposophical Quarterly*, London, 1975, Vol. 20, No. 4.

DESCOMBES, V., *Modern French Philosophy*, Cambridge, Cambridge University Press, 1980.

DINGLE, H., *Science at the Crossroads*, London, Martin Brian O'Keeffe, 1972.

DRAKE, S., *Discoveries and Opinions of Galileo*, New York, Doubleday Anchor, 1957.

DRAKE, S. and KOWAL, C. T., 'Galileo's Sighting of Neptune', *Scientific American*, 1980, Vol. 243, No. 6, pp. 52–9.

DREYER, J. L. E., *A History of Astronomy from Thales to Kepler*, New York, Dover, 1953.

DREYER, J. L. E., *Tycho Brahe: A Picture of Scientific Life and Work in the Sixteenth Century*, Gloucester, Mass., Peter Smith, 1977.

EATHER, R.H., *Majestic Lights – The Aurora in Science, History, and the Arts*, Washington D.C., American Geophysical Union, 1980.

EDWARDS, O., *A New Chronology of the Gospels*, London, Floris Books, 1972.

FAGAN, C., *Zodiacs Old and New*, London, Anscombe, 1951.

FAHIE, J. J., *Galileo: His Life and Work*, London, John Murray, 1903.

FARBER, M. (ed.), *Philosophical Essays in Memory of Edmund Husserl*, Cambridge, Mass., Harvard University Press, 1940.

FIRSOFF, V. A., *The Interior Planets*, Edinburgh, Oliver & Boyd, 1968.

FIRSOFF, V. A., *The Old Moon and the New*, London, Sidgwick & Jackson, 1969.

FRANCIS, P., *The Planets*, Harmondsworth, Pelican, 1981.

FRENCH, B., *The Moon Book*, Harmondsworth, Penguin, 1977.

FYFE, A., *Moon and Plant*, Arlesheim, Switzerland, Society for Cancer Research, 1967.

Selected Bibliography * 227

FYFE, A., 'The Signature of the Planet Mercury in Plants', *British Homeopathic Journal*, Oct. 1973, Jan. 1974, April 1974.

GOETHE, J. W., 'Nature – An Essay in Aphorisms', *Readings in Goethean Science*, Wyoming, Rhode Island, Bio-Dynamic Literature, 1978.

GOETHE, J. W., *Theory of Colours*, Cambridge, Mass., MIT Press, 1980.

GREENLER, R., *Rainbows, Halos and Glories*, Cambridge, Cambridge University Press, 1980.

HEATH, SIR T., *Aristarchus of Samos: The Ancient Copernicus*, New York, Dover, 1960.

HEIDE, F., *Meteorites*, Chicago, University of Chicago Press, 1964.

HERBERT, A. P., *A Better Sky*, London, Methuen, 1944.

HODSON, F. R. (ed.), *The Place of Astronomy in the Ancient World*, Oxford, Oxford University Press, 1974.

HOSKIN, M., *Stellar Astronomy*, Chalfont St Giles, Science History Publications, 1982.

HUMPHREYS, C. J. and WADDINGTON, W. G., 'Dating the Crucifixion', *Nature*, December 1983, vol. 306, no. 5945, pp. 743–6.

JAMES, M. R., *The Apocryphal New Testament*, Oxford, Clarendon, 1953.

JONES, K.G., *The Search for the Nebulae*, Chalfont St Giles, Alpha Academic, Science History Publications, 1975.

JULIAN, 'Hymn to King Helios Dedicated to Sallust', in *The Works of the Emperor Julian*, Harvard University Press, 1954.

JUNG, C. G., *The Interpretation of Nature and the Psyche*, London, Routledge & Kegan Paul, 1955.

JUNG, C. G., *Synchronicity: An Acausal Connecting Principle*, London, Routledge & Kegan Paul, 1972.

KAHN, P. G. K. and POMPEA, S. M., 'Nautiloid Growth Rhythms and Dynamical Evolution of the Earth-Moon System', *Nature*, 1978, vol. 275, no. 5681, pp. 606–11.

KAUFMAN, L. and ROCK, I., 'The Moon Illusion', *Scientific American*, 1962, vol. 207, no. 1, pp. 120–30.

KEPLER, J., *Kepler's Somnium: The Dream or Posthumous Work on Lunar Astronomy*, Madison, Wisconsin, Univeristy of Wisconsin Press, 1967.

KING, H., *The History of the Telescope*, New York, Dover, 1979.

KOLISKO, L., *Workings of the Stars in Earthly Substances*, Stuttgart, Orient-Occident Verlag, 1928.

KONNEN, G. P. and MEEUS, J., 'Triple Conjunctions: Twins and Triplets', *Journal of the British Astronomical Association*, 1982, vol. 93, no. 1, pp. 20–4.

LEHRS, E., *Man or Matter*, London, Faber & Faber, 1958.

LEVITT, I.M., 'Moon Illusion', *Sky and Telescope*, 1952, vol. 11, no. 6, pp. 135–6.

LEVITT, I.M., 'Mars Clock and Calendar', *Sky and Telescope*, 1954, vol. 13, no. 7, pp. 216–17.

LEY, W., *Watchers of the Skies*, London, Sidgwick & Jackson, 1964.

LOCKYER, N., *The Dawn of Astronomy*, Cambridge, Mass., MIT Press, 1973.

LODGE, SIR O., *Pioneers of Science*, London, Macmillan, 1926.

LOVELL, B., *In the Centre of Immensities*, St Albans, Granada, 1980.

LOWELL, P., *Mars*, USA, History of Astronomy Reprints, 1978.

LUDOVICI, L. J., *Seeing Near and Seeing Far*, London, John Baker, 1966.

LUM, P., *The Stars in our Heaven: Myths and Fables*, London, Thames & Hudson, undated.

MCDONNELL, J. A. M., 'The ESA Giotto Comet Halley Mission', *Yearbook of Astronomy*, London, Sidgwick & Jackson, 1983.

MACKENZIE, D., *Myths of Babylonia and Assyria*, London, Gresham, undated.

MALONEY, T., *The Sky at Night*, London, New English Library, 1963.

MANILIUS, M., *The Astronomica*, Cambridge, Mass., Harvard University Press, 1977.

MANN, I. and PIRIE, A., *The Science of Seeing*, Harmondsworth, Pelican, 1946.

MARTIN, M. E., and MENZEL, D. H., *The Friendly Stars*, New York, Dover, 1964.

MEEUS, J., 'Compact Planetary Groupings', *Sky and Telescope*, 1961, vol. 22, no. 6, pp. 320–1.

MEEUS, J., 'Extreme Perigees and Apogees of the Moon', *Sky and Telescope*, 1981, vol. 62, no. 2, pp. 110–11.

MEEUS, J., 'The Frequency of Total and Annular Solar Eclipses for a Given Place', *Journal of the British Astronomical Association*, 1982, vol. 92, no. 3, pp. 124–6.

MEEUS, J., *Astronomical Tables of the Sun, Moon, and Planets*, Richmond, Virginia, Willman-Bell, 1983.

MEEUS, J., and MUCKE, H., *Canon of Lunar Eclipses* – *2002 to* +*2526*, Vienna, Astronomical Büro, 1979.

MEEUS, J., GROSJEAN, C., and VANDERLEEN, W., *Canon of Solar Eclipses*, Oxford, Pergamon Press, 1966.

MEEUS, J. and GOFFIN, E., 'Transits of Earth as seen from Mars', *Journal of the British Astronomical Association*, 1983, vol. 93, no. 3, pp. 120–3.

MINNAERT, M., *The Nature of Light and Colour in the Open Air*, New York, Dover, 1954.

NEEDLEMAN, J., *A Sense of the Cosmos*, New York, Doubleday, 1975.

NEUGEBAUER, O., *The Exact Sciences in Antiquity*, New York, Dover, 1969.

NEUGEBAUER, O., *Astronomy and History* – *Selected Essays*, New York, Springer-Verlag, 1983.

Norton's Star Atlas, Edinburgh, Gall & Inglis, 1978.

OBERG, J., 'The New Case Against Extraterrestrial Civilizations', *Yearbook of Astronomy*, London, Sidgwick & Jackson, 1981.

OLCOTT, W. T., *Star Lore of All Ages*, London, Putnam's, 1911.

OLSON, R., 'Giotto's Portrait of Halley's Comet', *Scientific American*, 1979, vol. 240, no. 5, pp. 160–70.

O'NEIL, W. M., *Time and the Calendars*, Sydney, Sydney University Press, 1975.

OTTEWELL, G., *The Astronomical Companion*, Greenville, South Carolina, Furman University, 1979.

PANNEKOEK, A., *Periodicities in Lunar Eclipses*, Astronomical Institute of Amsterdam University, 1951, Circular no. 2.

PANNEKOEK, A., *A History of Astronomy*, London, Allen & Unwin, 1961.

PELTIER, L., *Starlight Nights*, Cambridge, Mass., Sky Publishing Corporation, 1965.

PERELMAN, Y., *Astronomy for Entertainment*, Moscow, Foreign Languages Publishing House, 1958.

PICKERING, W. H., 'The Time Relations of Astronomy and Geology', *Popular Astronomy*, 1919, vol. 27, no. 8.

PIRENNE, M. H., *Optics, Painting and Photography*, Cambridge, Cambridge University Press, 1970.

PIRENNE, M. H., *Vision and the Eye*, London, Chapman & Hall, 1971.

PLATO, *Timaeus and Critias*, Harmondsworth, Penguin, 1971.

PLATO, *Republic*, London, Macmillan, 1907.

PLUTARCH, 'On the Face in the Round of the Moon', in *The Origins of Scientific Thought* by Giorgio de Santillana, Mentor, New York, 1961.

POWELL, R., and TREADGOLD, P., *The Sidereal Zodiac*, London, Anthroposophical Publications, Temple Lodge Press, 1979.

PTOLEMY, *Tetrabiblos*, Slough, Foulsham, 1917.

ROSEN, E., *The Naming of the Telescope*, New York, Henry Schuman, 1947.

SANTILLANA, G., *Hamlet's Mill*, London, Macmillan, 1970.

SARDAR, Z., 'The Astronomy of Ramadan', *New Scientist*, 1982, vol. 94, no. 1311.

SARTON, G., *Ancient Science and Modern Civilisation*, Lincoln, University of Nebraska Press, 1954.

SCHARN, R., 'Extraterrestrial Beings Don't Exist', *Sky and Telescope*, 1981, vol. 62, no. 3, p. 207.

SEARS, D. W., *The Nature and Origin of Meteorites*, Bristol, Adam Hilger, 1978.

SHKLOVSKY, I., 'Is Life on Earth Unique?' *A Yearbook of Astronomy*, London, Sidgwick & Jackson, 1980.

SMART, W. M., *Some Famous Stars*, London, Longmans, Green, 1950.

SMITH, A., *The Seasons*, Harmondsworth, Pelican, 1973.

SPENCER JONES, H., *General Astronomy*, London, Edward Arnold, 1934.

SPENCER JONES, H., *Life on Other Worlds*, London, English Universities Press, 1952.

STEINER, R., *Truth and Knowledge*, Blauvelt, New York 10913, Steinerbooks, 1981.

STEINER, R., *Goethe the Scientist*, New York, Anthroposophic Press, 1950.

STEINER, R., *A Theory of Knowledge Implicit in Goethe's World Conception*, New York, Anthroposophic Press, 1978.

STEPHENSON, R., 'Historical Eclipses', *Scientific American*, 1982, vol. 247, no. 4, pp. 154–63.

STEPHENSON, F. R., and CLARK, D. H., *Applications of Early Astronomical Records*, Bristol, Adam Hilger, 1978.

THORNDIKE, L., *The Sphere of Sacrobosco and its Commentators*, Chicago, University of Chicago Press, 1949.

TRICKER, R. A. R., *The Paths of the Planets*, London, Mills & Boon, 1967.

TURNER D. and HAZELETT, R., *The Einstein Myth and the Ives Papers – A Counter-revolution in Physics*, Old Greenwich, Connecticut, Devin Adair, 1979.

UNGER, G., *'Ueber die Sagenannte Vertauschung von Merkur und Venus'*, Goetheanum, Dornach, Switzerland, *Mathematisch-Physicalische Korrespondenz*, 1979, no. 114.

WAERDEN, B. L. VAN DER, *Science Awakening II – The Birth of Astronomy*, Leyden, Noordhoff International Publishing, 1974.

WALKER, J. (ed.), 'Light from the Sky', *Readings from Scientific American*, San Francisco, California, Freeman, 1980.

WARNER, D. J., *The Sky Explored – Celestial Cartography 1500–1800*, New York, Alan R. Liss, 1979; Amsterdam, Theatrum Orbis Terrarum, 1979.

WASHBURN, M., *Mars at Last!*, London, Sphere Books, 1979.

WATERFIELD, R., *The Revolving Heavens*, London, Duckworth, 1944.

WHITE, J., *The Birth and Rebirth of Pictorial Space*, London, Faber & Faber, 1972.

WHITMELL, C. T., 'The Moons of Mars', *Journal and Transactions*, Leeds Astronomical Society, 1903, no. 11.

YEOMANS, C., *The Comet Halley Handbook*, Pasadena, California, National Aeronautics and Space Admin., 1981.

Index

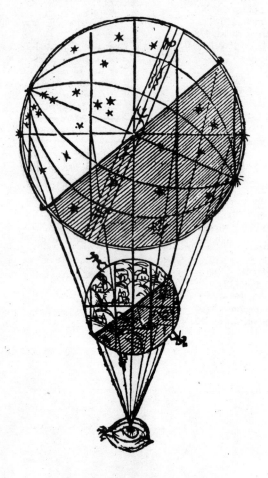

Eye, earth and cosmos in unity.
From Cosmographicus liber (1533) by
Peter Apian, professor of mathematics
at the university of Ingolstadt.